Cheating Monkeys and Citizen Bees

LEE DUGATKIN

Cheating Monkeys and Citizen Bees

The Nature of Cooperation in Animals and Humans

Harvard University Press
Cambridge, Massachusetts
London, England

Printed in the United States of America

First Harvard University Press paperback edition, 2000

Originally published by The Free Press, a division of
Simon & Schuster, Inc.

Library of Congress Cataloging-in-Publication Data
Dugatkin, Lee Alan, 1962–
Cheating monkeys and citizen bees : the nature of cooperation in animals and humans / Lee Dugatkin.
p. cm.
Includes bibliographical references and index.
ISBN 0-674-00167-2
1. Social behavior in animals. 2. Animal behavior—Evolution. 3. Cooperativeness (Psychology). I. Title.
QL775.D84 1999
591.56—dc21 98-27768

To my father, Harry,

on his eightieth birthday.

Mazel Tov.

Contents

	Preface	**ix**
INTRODUCTION	*The Four Paths to Cooperation*	**1**
1	*All in the Family*	**39**
2	*One Good Turn Deserves Another*	**77**
3	*What's in It for Me?*	**109**
4	*For the Good of Others?*	**135**
CONCLUSION	*Possibilities and Pitfalls*	**165**
	Notes	**175**
	Index	**201**

Preface

For the first five years that I studied the evolution of behavior, I was what I now refer to as a "for the love of science" scientist. That is, I was absolutely fascinated with evolutionary biology, and it was my love for the subject that drove me to do what I did. I was not concerned about what direct, practical impact my work would have on people; I was simply convinced that it was inherently interesting. In fact, it went further than that. During this stage of my intellectual development, I actually held those scientists who directed their work toward helping people in contempt. I viewed them as intellectual prostitutes, who sold their souls and were driven by the mundane, everyday wants and needs of the masses rather than by the sheer beauty and excitement that made up the heart of science. I am ashamed that I ever held such a view, but I did.

I began studying the evolution of cooperation in animals about twelve years ago. I found the topic fascinating, and despite its clear implications for humans, I was unconcerned about what my work would tell us about how to make people more cooperative. If pressed, I would justify my work to friends and family by showing them how one *could* use what I do to understand human behavior, but that was not what drove me to do my work, it was merely a happenstance. I would have been just as happy studying some area of evolution that did not so easily lend itself to human analogies. Furthermore, the ideas I did have on how my work on animal cooperation might apply to human cooperation were at best fuzzy and undeveloped.

After operating in the "for the love of science" mode for

about five years, I met my future wife, Dana, and soon our wonderful little boy, Aaron, was on the scene. In addition to the huge changes they brought about in my everyday life, Dana and Aaron had a profound effect on how I thought about science. Perhaps cooperation became more critical in my personal life. For whatever reason, I became more and more convinced that my work had relevance to those who were scientists by training. When I looked in Aaron's trusting eyes, I realized it was only a matter of time before he asked me what science was. I decided that the best way for me to answer him would be to describe the beauties of scientific investigation and also to identify how it can have human consequence. As a start, I intend one day to share with him my thoughts on how ideas emerging from the study of animal cooperation might facilitate human sociality. This book is for him.

I am greatly indebted to my editor at the Free Press, Stephen Morrow. Stephen had the unenviable task of taking on this book halfway through the editorial process, but he quickly began providing critical advice that had a significant impact on the final product. Susan Arellano worked with me on the early stages of the manuscript (including encouraging me to write the book in the first place), and her input helped set the tone for what followed.

I have talked about the evolution of cooperation with hundreds of colleagues over the last ten years, and while I can't name them all, I am obliged to mention three. My mentors Jerry Brown and David Sloan Wilson taught me almost everything I know, and I can never repay them for what they have given me in the way of both advice and friendship. Michael Mesterton-Gibbons was absolutely integral in helping me to develop mathematical models of the evolution of cooperation, and I look forward to our future work in this area.

Without the love, encouragement, and help of my wife, Dana, this book would never have come to fruition. In addition to proofreading every chapter more times than I (or she) care to recall, Dana also was pivotal in helping shape the ideas in this book. She is an amazing lady. My parents, Harry and Marilyn, and my brother David have encouraged me through the years. They read through this book and provided me with much needed advice. Lastly, I thank my little boy, Aaron, for helping me keep things in perspective.

Cheating Monkeys and Citizen Bees

The Four Paths to Cooperation

It is true that certain living creatures, as bees and ants, live sociably one with another . . . [and] some men may desire to know why mankind cannot do the same. To which I answer:

First, that men are continually in competition for honor and dignity, which these creatures are not; and consequently amongst men there ariseth on that ground, envy and hatred, and finally war. . . .

Secondly, that amongst these creatures, the common good differeth not from the private. . . . But man, whose every joy consisteth in comparing himself with other men, can relish nothing but what is eminent.

Thirdly, that these creatures, having not, as man, the use of reason, do not see, nor think they see any fault, in the administration of their common business; whereas amongst men, there are very many, that thinks themselves wiser, or abler to govern the public, better than the rest; and these strive to reform and innovate . . . and thereby bring into distraction and civil war.

Fourthly, that these creatures, though they have some use of voice, in making knownst to each other their desires, and other affections; yet they want that art of words, by which some men can represent to others that which is good, in the likeness of evil; and the evil in the likeness of good. . . .

Fifthly, irrational creatures cannot distinguish between injury and damage; and therefore as long as they be at ease, they are not offended with their fellows; whereas man is then most troublesome, when he is most at ease; for then it is that he loves to . . . control the actions of them that governth the commonwealth.

Lastly, the agreement of these creatures is natural; that of men is by covenant only, which is artificial; and therefore it is no wonder if there be somewhat else required.

Writings of a biologist? A psychologist? No, instead these words come from *Leviathan*, the work of seventeenth-century philosopher and political scientist Thomas Hobbes. Was Hobbes right about animal and human cooperation? What, in fact, is the nature of cooperation?

Scientists who study human behavior argue about many things—in fact, they argue about everything from the minutiae of everyday life to grander questions regarding good and evil, the motivations underlying our actions, and whether we are driven primarily by genes, culture, or some intricate interaction of the two. One thing they do not disagree about, however, is whether humans are social creatures. We are, period. We know of no cases throughout history where large numbers of humans have intentionally lived outside the fabric of *some kind of* society. Anthropologists studying man's evolution find that sociality is our natural state.[1] Granted, the occasional hermit, living on his own in a cave for forty years, is found from time to time. Yet we know of such cases primarily because of their rarity and because they seem, in one way or another, unnatural to us—it is just not what we humans do. We simply are not designed, either physically or psychologically, to live as solitary creatures. There is a reason that solitary confinement is so dreaded by prisoners.

Within societies all across the planet, be they small nomadic groups of kin wandering through the grasslands or millions of unrelated individuals living in a metropolis, whether modern or prehistoric, cooperation is the glue that binds us together. It is difficult to even imagine a society in which cooperation, at some level or another, has not been integral. Certainly all groups have the requisite cheaters among them, and we will spend a great deal of time on such individuals: How should we punish them? How can we avoid interacting with them? Why do cheaters act the way they do? Nevertheless, without joint actions aimed at the production of something useful—that is, without cooperation—a society, almost by definition, is bound to crumble.

One possible gauge for estimating just how important cooperation is in all aspects of life is to recognize that we teach it to our children as early as they are able to grasp a story about others. Consider *The Little Red Hen,* a tale read by millions of parents to their toddlers.

> One morning the Little Red Hen was pecking in the barnyard when she came across some grains of wheat. "How nice it would be to plant the grains and grow some wheat and bake some bread." So the Little Red Hen gathered up the grains of wheat and said: "Who will help me plant this wheat?"
>
> "Not I," quacked the Duck. "Not I," meowed the Cat. "Not I," grunted the Pig. So the Little Red Hen planted the grains of wheat all by herself.

The Little Red Hen then solicits the duck, the cat, and the pig to help her cut the wheat, bring it to the mill, and bake the bread; each time she is answered with the same chorus of "Not I." Finally, the Little Red Hen

> . . . baked a warm and tasty loaf of bread.
>
> "Now, who will help me eat the bread?" called the Little Red Hen. "I will," quacked the Duck. "I will," meowed the Cat. "I will,"

> grunted the Pig. "Oh, no you won't," said the Little Red Hen. "I planted the wheat. I cut the wheat. I took it to the mill to be ground into flour. And I baked this bread without any help from the three of you!"
>
> Then the Little Red Hen took a bit of fresh butter, sat down under a shady tree, and ate the loaf of bread . . . all by herself!

Now, while it is true that in the end the Little Red Hen bakes her bread without the help of others, the message of the story really centers on the inaction of the duck, cat, and pig. From this trio, we learn that if you fail to contribute to the production of a resource that is potentially available to all group members—in this case a loaf of bread—you simply will not be allowed to get the resource for free and thereby parasitize the efforts of others. My four-year-old son doesn't quite articulate it as such, but that is the bottom-line message. Of course, we all know of many instances in which this ideal is not upheld, but the point is that we teach our children the ideal, because it reflects what so many of us believe about human nature, or at the very least what we want to believe about human nature.

We read our kids *The Little Red Hen* because we cannot avoid the shadow of cooperation in virtually everything we do, or at least attempt to do. Consider just a few examples for the moment; we will delve into many more in later chapters.

During my drive in to work, I often listen to National Public Radio. I recognize that public radio listeners need to face up to a fact: if you aren't going to be forced to listen to commercials about the fish food sale at your local pet store, somebody is going to have to pay the bills for broadcasting this stuff. And that means that the dreaded words you will be inundated with for one solid week—*pledge drive*—are always just around the corner.

The issue of cooperation arises for all public radio listeners. What should we do when pledge week comes around and everyone at the station, from the president to the assistant director of marketing, takes a stab at convincing us that we should pitch in and contribute? Although not glaringly obvious, there is a true dilemma here in terms of whether one should cooperate or not. After all, I can easily convince myself that if I don't contribute, enough people will, and the station will continue to operate. Not only could I convince myself of this, it would in all likelihood be true. So, perhaps I wouldn't cooperate and send in my pledge. But if this is true for me, it is true for *every* public radio listener. What if everyone decided not to contribute because others will? The answer is clear—no more public radio. So, while public radio is not relying on any *single* person to contribute, thus providing a temptation not to contribute (you get the product free), if no one cooperates, everyone loses.[2]

Enough people cooperate and contribute to keep public radio alive and well, but there are many other similar dilemmas where cooperation is not the inevitable outcome. Consider the brown-outs that are notorious in hot parts of the country during scorching summer days. Out of the blue, the electricity in large areas shuts down. Why? Could air conditioner owners, each deciding whether or not to cooperate, be responsible? Each owner believes that if she keeps her air conditioner on the maximum setting, that alone will not overload the system—so, to keep comfortable, that is precisely what she does. But if this holds true for one person, it is true for that person's neighbor, her neighbor's neighbor, and so on, and when everyone comes to the same decision to crank up the air conditioner, we get a brown-out. So the lack of cooperation imparts a price we all pay, like it or not. But it need not always be that way. Josh Weiner and Tabitha Doescher, for example, found that in a survey of utility customers, individuals were more likely to install control devices on their cooling units *when they thought others would act in a similar fashion.*[3]

Returning to the public radio example, imagine that the story that airs after the pledge drive focuses on the nuclear arms race between the United States and the former Soviet Union. Hard as it may seem to swallow, this is yet another version of the same cooperation problem we just faced, except that in this case, countries are the decision makers and there is a lot more at stake. Let us assume, for the time being, that both parties (the U.S. and the former Soviet Union) would prefer to have no nuclear weapons on the planet. As such, mutual cooperation is the desired goal. But why, then, is it so hard to get there? Why were both the United States and the Soviet Union using strategies like "mutual assured destruction" (any nuclear attack would lead to the destruction of all parties) for so long? One possibility is that neither side could disarm its nuclear weapons first, because cooperation by one party and noncooperation by the other would be an intolerable position for the cooperator. Mutual cooperation is again best for everyone, but the temptation to cheat (not cooperate) is very strong because of the consequences of unilateral action. Clearly, whether parties eventually cooperate in this instance and how they choose to do so has social and political consequences.

Philosophers and psychologists, economists and biologists have been making conjectures on the nature of human cooperation for as long as those professions have been around. Let's take just a few moments to look at some historical aspects of the study of cooperation, to put this subject in some perspective.

Philosophers tend to believe that humans are either naturally social and cooperative or naturally antisocial and uncooperative. Aristotle fell in the first camp and was quite adamant about it: "Man is by nature a social creature: an individual who is unsocial naturally and not accidentally is either beneath our notice or more than human. Society is something in nature that precedes the individual. Anyone who either cannot lead the common life or is so self-sufficient as not to need to and therefore does not partake of society is either a beast or a god" (*Politics,* 328 B.C.).

Aristotle believed that the function of man was to lead a life of reason, because it was this alone that separated us from the beasts.[4] On the other hand, there have been those who believed mankind was capable of cooperation, but that individuals would only cooperate if some external force saw to it. *The Sayings of the Fathers,* a fourth-century compilation of advice on religion and social affairs, notes ominously: "Pray for the welfare of the government, since but for the fear thereof men would swallow each other alive."

The two most famous combatants to square off on the issue of whether people are by nature cooperative or not were undoubtedly John Locke and Thomas Hobbes, British philosophers and political scientists of the seventeenth century. Both Hobbes and Locke believed humans to be capable of cooperation; the question they differed on was whether we are *innately* cooperative. In essence the debate was on the most fundamental of all questions: are we naturally good or evil? As such, the debate was often cast in terms of whether war was the default state of human behavior or an aberration and whether some governmental unit was needed to enforce cooperation among civilians.

In *Two Treatises of Government* (1690), Locke sets out his rather far-sighted views that man is born free, that governments are not divinely appointed but rather a form of contract that can be broken by the people at any time, and that man is basically good and naturally inclined toward cooperation rather than destruction (via war): "And here we have the plain difference between the state of nature and the state of war, which however some men have confounded, are as distant as a state of peace, mutual assistance and preservation: and a state of enmity, malice, violence and mutual destruction are from one another."

Thomas Hobbes, whose *Leviathan* predates Locke's work by half a century, saw things in quite a different light. Mankind in its natural state, without a powerful government to keep citizens in line, was, Hobbes believed, "in a condition called war; and such a war as is of every man against every man." Government's function, according to Hobbes, was to enforce cooperation among its constituents, and he believed that the covenants formed between the governors and governed were based on force. In Hobbes's eyes, "covenants, without the sword, are but words and of no strength to secure man at all." Hobbes went as far as codifying this view in his *Fundamental Law of Nature,* which includes the supposition that "every man has a right to every thing: even to another's body." Without governments to see that cooperation exists among constituents, Hobbes believed, life would be "solitary, poor, nasty, brutish and short." An interesting corollary to this "fundamental law" was that man was aware of his condition. Without government, people simply refused to invest in any public goods, since such goods would no doubt be usurped by others. The existence of government was not only needed to bring about cooperation per se, but was also the only means for people to procure investments and common goods. Somewhat surprisingly, despite his rather gloomy view of human nature, Hobbes believed that animals could be quite cooperative, by instinct (see his words at the beginning of this chapter).

The debate made famous by Locke and Hobbes still burns strong today. Are we inherently good? Do we tend toward cooperation, if given the chance? Today, however, the proponents are not particularly interested in the facts about cooperation; rather, they focus on how to make the opposing viewpoint seem foolish. Religions differ on our "fundamental" nature (good? evil? neither? both?), and even political parties can often be divided according to whether they believe that we are fundamentally good (as most liberals believe) or that we are neither good nor bad, but learn to be these things (as most conservatives believe). Resolving the differences between religions and different political parties is probably impossible and perhaps not even advisable. But given the tremendous advantages of actually understanding the very core of our behavioral repertoire, we must continue our investigation into this matter. While philosophy has certainly proved useful in illuminating the *questions,* perhaps it is not the ideal place to look for the *answers* about our cooperative tendencies (or lack thereof). But economics might be.

Economists and business people are, not surprisingly, acutely interested in cooperation, albeit for somewhat different reasons than philosophers. The approach taken in economics centers on the notion of the "rational man." Most, but certainly not all, economists argue that people, one way or another, assess the costs and benefits of taking some action that has economic consequences and use a simple rule of thumb. If the benefits of the action outweigh the costs, people undertake it; otherwise they do not—hence the decision is considered rational. Extending this idea to cooperative behavior, people cooperate with one another when it is in their own interest to do so (the benefits are greater than the costs) and refuse to cooperate in other situations.

This notion of the rational man has its roots deep in the history of economics and can be traced back at least as far as Adam Smith's *The Wealth of Nations,* wherein he laid out his famous "invisible hand" theory. The "invisible hand" ensures that the self-interest of each party translates into a well-oiled economy.

> It is not from the benevolence of the butcher, the brewer and the baker that we expect our dinner, but from regard to their own self-interest. We address ourselves not to their humanity but to their self-love and never talk to them of our own necessities but of their advantages. Nobody but a beggar chooses to depend chiefly on the benevolence of his fellow citizens.[5]

Cooperation, Adam Smith believed, is a natural result of individuals trying to maximize their profit. When cooperation fails to serve this function, people don't cooperate. This approach underlies the literally thousands of studies on the subject performed in the area of social psychology. In most of these economics-based psychology experiments, a group of people are assembled in a psychology lab and given instructions about a "game" that they are about to play. Typically, individuals are given a small amount of money and told that if enough people are willing to contribute a portion of the money they originally received, everyone in the group will get a relatively large bonus (notice the similarity to the public radio dilemma). For example, after being given $10 each, a group of eight subjects will be told that if four of them contribute $5 each, then everyone in the group, *contributors and noncontributors alike,* will be given a $10 bonus; but if fewer than four players contribute, no bonus will be given *and those who contributed will be out their $5.* Social psychologists then examine how such factors as group size and the ability of subjects to talk over the matter before deciding what to do affect whether the "cooperative bonus" is obtained and who is willing to contribute. Underlying all the manipulations, social psychologists are generally interested in whether people can figure out the rational choice that would bring them the most money.

The notion that humans are inherently selfish, cooperating only when it is in their economic best interest to do so, has certainly not gone unchallenged.[6] In addition to the anecdotal cases we can all come up with, there have been numerous experimental attempts to determine whether we are continually acting as cost/benefit detectors, basing our decisions on this one element alone, or whether the story is more complex. Linnda Caporael and her colleagues, for instance, have argued that the rational man theory of cooperation (or the "egoistic incentive" theory, as they refer to it) is largely untested and based more on cultural beliefs than empirical evidence. In an attempt to test this, they examined whether people would cooperate with each other in the *absence* of economic incentives to do so. Caporael's team found that people indeed do cooperate without such incentives and that the participants themselves often cite "group welfare" as a primary cause for such "irrational" decisions.

In fact, the social psychology notion that people need to be tested to see if their acts are based solely on selfishness has its roots in the writings of the ancient prophets and is essentially what the story of Job is all about.[7] In this parable, the Devil approaches God and puts forth the following argument: The only reason Job loves You is that You (God) have given Job everything he needs, and more. Take away that incentive, says the Devil, and You will see just how quickly Job reconsiders his devotion. God does just that, and of course Job, though a bit confused about what is going on, remains loving and devout in the absence of any economic incentives. How common the attitudes of Job and Caporael's subjects are remains a matter of debate, but they certainly pose a respectable alternative to standard economic thinking about cooperation.

Despite challenges to its universality, the rational man theory, because of its predictive power, has a very strong foothold not only in its home discipline of economics but in political science, psychology, anthropology, and even biology. There are, however, a few practical problems with this approach. First, there is the question of timescale. When we say that people will cooperate if the benefits of cooperation outweigh the costs, do we mean with respect to the immediate consequences of an action or to the longer-term consequences? If cooperating with you today harnesses me with some new cost, but five years down the road I receive a benefit as a result of my behavior, is it rational for me to perform the action now? Second, and in many ways more problematic, is how to incorporate many different currencies into an economic decision about whether to cooperate (in technical terms, this is a problem with utility functions[8]). If cooperating with someone gets me three apples but not cooperating gets me an orange and $1, how do I decide what to do? Which is worth more (that is, which provides me with greater satisfaction)? This is the precise reason we have a monetary system with well-defined units, but translating across apples and oranges to dollars (or whatever currency you are using) is not always straightforward, and it is not clear that people even attempt to do this, in any real way, when deciding whether or not to make a deal.[9]

The philosophical, psychological, and economic approaches to the study of cooperation are all illuminating in their own ways. But they fail to address a fundamental problem with respect to cooperative behavior: How could cooperation persist *over long periods of time,* when there seem to be so many ways that individuals who don't cooperate can circumvent the system? Over the course of many generations, why don't we see cheaters (those who fail to cooperate) slowly increase in frequency, replacing their cooperating peers? If you can benefit from others cooperating, why should you ever cooperate? Cheaters obtain the resources that cooperators obtain but don't pay the costs of cooperation.

To fully address these issues, we must now turn to the science of evolutionary biology. Here, using natural selection as our guidepost, we can examine the various means by which cooperative strategies can evolve over vast stretches of time, as well as the reasons cooperation often fails to manifest itself in certain conditions. It is with the techniques developed in evolutionary biology, and more specifically behavioral ecology, that we will pursue our study of cooperation throughout this book. As we shall see, evolutionary biology and behavioral ecology are uniquely suited to address questions surrounding the existence and maintenance of cooperation, as the conceptual and mathematical tools have already been developed (in other contexts) to address the evolution of cooperation. Behavioral ecology is a discipline whose primary function is to study the evolution of social behavior, and so it is the area of biology best suited to examine cooperation.

The work compiled by behavioral ecologists over the last twenty-five years or so is an untapped treasure—not just as a checklist of cases of cooperation, but also as a conceptual spawning ground for ideas on how to promote human cooperation. Before examining the nuts and bolts of an evolutionary approach to the study of cooperation, let me put forth my hypothesis regarding precisely *how the study of evolution and animal behavior can be used to foster and enhance cooperation in humans.*

What we can learn about human cooperation by studying nonhumans begins with this assumption: We are much more cognitively sophisticated than animals. Despite considerable work done recently (some of it coming from my own lab) suggesting that animals are behaviorally far more complex than we have given them credit for, everything is relative. Animals can be much more sophisticated than we previously believed and still much less cognitively advanced than humans. It is worth emphasizing this point, because some have argued that one reason we can use animal examples to understand human nature is that some animals possess a rudimentary form of morality. It would require a separate book to elaborate on why I do not hold that view, but I believe animal work on cooperation can be profoundly useful for fostering human sociality nonetheless.

My hypothesis is that animal cooperation shows us what to expect when the complex web of human social networks, as well as the laws and norms found in all human societies, are absent, and so these studies act as a sort of baseline from which to operate. Animals show us a stripped-down version of what behavior in a given circumstance would look like without moral will and freedom. Only with this understanding of what a particular behavior looks like outside the context of some moral code can we use human morality to focus on and foster cooperation in our species.

Studies on cooperation in animals can be used to help us better understand and promote human cooperation in two ways. Both begin with a thorough search for common factors that have been found in many cases of animal cooperation. Once such factors are uncovered, one thing we can do is to identify areas in which we see failures in human cooperation, and then use our knowledge of animal cooperation to add critical factors to the human scenario. This might be referred to as the "missing element" approach. For example, if we have found that cooperation in animals is common when individuals frequently interact, we have identified a factor promoting animal cooperation. We can then examine how we might add this factor to the human scenario with which we are concerned. A second way of using animal studies is to identify human behavioral scenarios containing some of the factors that we know tend to favor cooperation in animals, and to use various techniques to amplify these factors and hence enhance the probability of cooperation. This might be thought of as the "nudge it over the top" approach.

Throughout this book, we shall constantly return to both of the above approaches. Put simply, the evolutionary work on cooperation in animals is a virtually untapped treasure chest in terms of understanding human cooperation. To overlook it would be a shame and perhaps much worse than that.

Before moving on to a few more detailed cases of how work on cooperation in animals can be used to foster human cooperation, I want to address a broad-based objection to the sort of approach I am advocating. One could argue that we humans are so different from animals that any attempt to study a phenomenon in animals is completely irrelevant with respect to that phenomenon in humans. There are literally thousands of examples one could give to refute this argument, but consider just one for the purpose of illustration.

Since the time of the Spanish explorer Ponce de Leon, people have been fascinated with the nature of getting old (in biological terms, the process of senescence). Our interest in why fruit flies get old, however, is probably a bit less heartfelt. Yet all across the world there are laboratories whose sole function is to study aging in fruit flies. This work has proven very productive and has shed a great deal of light on why humans, as well as fruit flies, get old and what we might do to affect the process of aging. Fruit flies are a model species, a means to reach an end, and they have proved quite useful in helping us better understand ourselves.

The reason that fruit flies help us understand aging in general is that they contain genes that are similar to the genes we believe underlie aging in humans. But many organisms have some genes in common with us. In fact, one unlikely candidate for studying aging in humans has proved a veritable gold mine—yeast! About 31 percent of the yeast genome have counterparts (called homologs) in humans.[10] David Sinclair and his colleagues at the Massachusetts Institute of Technology, armed with this information, have studied the *sgs1* gene in yeast, which has a counterpart labeled *WRN* in humans, to better understand the aging process. Mutations in *WRN* are known to result in Werner's syndrome in humans, producing premature aging. Likewise, Sinclair and his colleagues found that *sgs1* mutations caused premature aging in yeast cells.[11] More than that, however, they were also able to understand some of the molecular details of how *sgs1* caused aging. Studying yeast genes paid off by providing previously unknown details about the aging process in both yeast and humans. Similar arguments have been made with respect to yeast experiments and our fundamental understanding of the biology of cancer. Humans can be fundamentally different from flies and yeast and yet still share certain commonalities. To argue that this line of reasoning might be true for traits like aging but probably not for behavior is not grounded in any logic.

To guide us in examining how animal studies can shed light on human cooperation, we need a better understanding of how cooperation manifests itself in nonhumans. So let's take a look at some examples. Four different paths to cooperation have been outlined by behavioral ecologists (some routes being more controversial than others). These paths go by various names, but here we shall begin by labeling them (1) family dynamics, (2) reciprocal transactions, (3) selfish teamwork, and (4) group altruism. These four pathways provide a framework for evolutionary thinking on cooperation and have been examined in some detail in nonhumans.

The First Path: Family Dynamics

In an open field somewhere in California, a group of ground squirrels peacefully go about their normal daily activities. Seemingly out of nowhere, a hawk begins its deadly dive from the air, targeting the squirrels for its next meal. Suddenly a piercing shriek echoes through the valley—the alarm call of one female squirrel. The field comes to life with squirrels making mad dashes toward their burrow or some other safe haven. Minutes later, when the hawk has clearly set its sights elsewhere, the squirrels slowly begin to resurface.

A puzzle emerges in this example. Why should an individual squirrel be the potential "sacrificial lamb"? After all, screaming alarm calls as loudly as possible must make you the single most obvious thing in the entire field. Why attract the hawk's attention and most likely make yourself its next snack? Why not let someone else take the risks? As with so many riddles in the field of evolution and animal behavior, the answer lies in the interaction between ecology and kinship.

The first step in unraveling the puzzle of why a female should volunteer to give a warning call lies in recognizing that the group of squirrels being warned of upcoming perils is not a random assortment of animals. Rather, these squirrels live in a society full of relatives, and for those who study the evolution of social behavior, that makes all the difference in the world. Your relatives, by definition, are more likely to carry the same genes that you possess than are a handful of strangers picked off the street. These genes are called "identical by descent" because the likelihood of sharing them is related to descent from some common ancestral relative. For example, the common ancestors among sisters are their mother and father, while those for cousins would be a grandmother and grandfather. What this means is that when a squirrel risks its own life to save the lives of its relatives, it is not being completely altruistic, because its relatives most likely carry the same genes that it does. Since sisters, for example, share (on average) 50 percent of their genes that are identical by descent, saving two sisters and losing your life is equivalent to an even swap, while saving three sisters and losing your life is a net positive for all parties involved! And if you are only *risking,* rather than necessarily *giving up* your life, you might do it if fewer relatives were around.

Yet, so far, we have only explained half the riddle—the half surrounding why *anyone* should give alarm calls. Now let's look at *who* gives a call. One perplexing element in our story is that sex seems to play a role in establishing who our cooperative sentinel is. Over and over again, we see that females, not males, put their necks on the line. Why? Again, that overriding theme in the evolution of behavior—kinship, this time mixed with choice of living venue—plays a role.

In many species of animals, males and females do not make the same decision about where to set up residence. Once an individual is mature enough to survive on its own, it is faced with a rather daunting decision: should it live near where it grew up or emigrate elsewhere? Despite what your teenager will tell you, this is not a straightforward decision. Living in the old neighborhood clearly has both costs (e.g., competition from local rivals) and benefits (e.g., security), as does moving to a new location. Furthermore, the costs and benefits can be very different depending on whether you are male or female, and in the animal kingdom we see every combination possible—males stay and females leave, males leave and females stay, both sexes leave when reaching maturity, and both sexes remain in the natal home area. Among squirrels, it is the males that leave home to set up shop elsewhere and the females that stay in their old stomping grounds.

Once males leave and females stay put, an interesting imbalance occurs. Females now find themselves surrounded by relatives, while males are in areas with complete strangers. This asymmetry in relatedness then favors dangerous alarm calling in females, since they will be assisting kin, but the same does not hold true for males. Blood, then, explains both why an individual should put itself in harm's way by calling out when a predator is sighted and who should be most likely to do so. If this is true, then an interesting prediction can be made. On the rare occasions when females are forced to emigrate to new groups, these newcomers, despite being female, should be among the least likely individuals to give alarm calls. Sure enough, that is precisely what we find.[12]

The Second Path: Reciprocal Transactions

Walk a bit along the beautiful, calm streams of the Northern Mountains of Trinidad, West Indies, and you will quickly see that one species common to these waters is the guppy. In many streams, the water is crystal clear and the behavior of guppies can be seen from the bank. If you sit there for a few hours, you will probably observe a pair of guppies breaking away from their group and approaching a dangerous predator. This alone is bizarre enough, given that the fish could just as easily have headed for cover, rather than toward this menace. But not only do some fish take the risks inherent in approaching a predator, they actually seem to return to their group and somehow pass on the information they just obtained.

Our risk-taking guppies are trapped in a dilemma. The temptation to cheat is always there—the best thing that could happen to guppy 1 is for guppy 2 to take the risks and pass the information on to the other guppy for free. But if both individuals opt to wait for the other to go out and do the dirty work, they may be worse off than had they just gone and done it as a team. Many other examples of this dilemma pervade everyday animal life.

Is there an escape from this dilemma? Can we identify any sort of cooperation in this example? The answer is yes, providing a *pair* of individuals finds themselves in a similar predicament *many times*—certainly plausible for both guppies. When partners are paired up many times, evolutionary theory predicts that individuals should use what is called the "tit-for-tat" strategy. Tit-for-tat is just another term for the "eye for an eye" rule found in Exodus. That is, start out cooperating with your partner, but if he cheats on you, cheat right back in the same manner he did. Despite having a brain not much bigger than a pinhead, guppies appear to use the tit-for-tat rule on their sorties toward potentially dangerous predators. Each fish keeps track of what the other is doing when both go out to examine the predator. Should one fish lag a little behind, the other fish slows down and makes sure that the distance does not become too great. To top it off, guppies genuinely prefer to spend their time hanging around other guppies who cooperated with them during their danger-filled sorties, presumably to be in their vicinity again, should the situation arise once more![13] If guppy predator inspection reminds you of guard duty in the army, you are not alone, but more on that after a few additional animal examples.

The Third Path: Selfish Teamwork

Watching lionesses hunt a gazelle on the plains of Africa is a savage, beautiful, and mesmerizing event. Lionesses don't just mindlessly chase after a gazelle; rather, the hunt is a masterpiece of coordinated action aimed at one result—a gazelle meal. Often one lioness will flush the gazelle and one or more of the others will chase it, or each hunter will come at the prey from a different angle, limiting maneuvering room for the gazelles. Such coordinated hunting is seen in other animals as well. But is it cooperation? Might it not be the case that each lioness strictly has her own interest, and that alone, at heart? Yes, but this does not detract from the fact that the hunt is clearly a coordinated action set to accomplish a particular goal (motives and psychology aside). In fact, this type of cooperation, one in which joint action is predicated only on the self-interest of all parties involved, may be the most common type of animal cooperation.[14]

The Fourth Path: Group Altruism

The Sonoran Desert is home to a fascinating species of ants named *Acromyrmex versicolor.* Although ant nests are usually initiated by sisters, *Acromyrmex* queens are unrelated. Despite not being relatives, queens in this species are friendly—no one attacks anyone else in the nest, all food is shared equally by everyone, and all the queens have about the same number of offspring. Yet in the midst of this cooperation fest, a riddle emerges: only one queen in the whole nest goes out and gets food for everyone. This sounds particularly odd because the underground nest is a pretty safe place to be, but going out into the desert is rather like stepping into a minefield. Any number of species out there view queens as juicy morsels, and many foraging ventures end in the demise of the forager. On top of all that, which queen ultimately becomes the group's food provider seems to be determined by chance—no one is coerced into taking the job.

How could such dramatic, "for the good of others" cooperative behavior ever come to be? If these queens are not related, why should any one of them accept the role of food gatherer with all of its risks, when the booty that comes in is divided between all group members equally and ultimately leads to all nestmates producing similar numbers of offspring? The answer to these questions lies in changing perspectives. Rather than viewing this strictly in terms of the costs and benefits to the *individual* who is the gatherer, one must expand the notion of costs and benefits to the group. This perspective is called *group selection.*

The basic premise of group selection is quite simple. When considering any behavior, one must examine the effect the behavior has on the individual undertaking it *and* those around it. If the behavior is beneficial to all involved, no obstacles exist to its evolution. If the behavior has negative effects on all parties involved, then such behavior disappears very quickly. But what about the queens in our example? Here we have a case where the behavior (food gathering) has a negative impact on the forager but a positive impact on the group. When should we expect this sort of cooperation to persist? The answer, though it can be couched in complex mathematics, is also just plain intuitive—when the group-level positive effects outweigh the individual-level negative effects.

Despite costs to the queen, who takes all the risks associated with sorties from the safety of the nest, the group-level benefits of this action are so great that the behavior persists. To see why, we need yet one more piece of information about *Acromyrmex versicolor.* And that is that once queens raise their offspring, everyone comes up to the surface and what amounts to all-out warfare between individuals in different nests occurs. Only one group will survive. The odds of surviving are directly related to the number of fighters each group has, and that is simply connected to how much food queens had and how many offspring they could produce.

So, in the extreme, imagine nests in which no one would go out and forage and everyone would try to produce just whatever progeny they could from their stored body fat. Such a nest would surely perish in the nastiness to come, and hence the group-level benefits of having a specialized food gatherer outweigh the costs to the individual undertaking the action.[15]

. . .

Examples such as these, in conjunction with evolutionary theory, serve as a starting point for our discussion of how cooperation in animals can better help us foster cooperation in humans. We will journey into beehives and naked mole-rat colonies to see a single female producing all the offspring for an entire group and a vast array of others who seem intent on cooperating with each other to help such a queen. Fish (and worms) switch sexes in order to divide up reproduction in a cooperative manner. Impalas cooperate with each other by pulling parasites off hard-to-reach places on the bodies of their friends to make them more hygienic. Mongooses take turns baby-sitting for each other. What will tie together these stories of cooperation in animals, however, is what they can and cannot tell us about cooperation in humans. In this context, we are interested not only in how natural selection has shaped a vast array of cooperative acts in a very diverse group of organisms, but also in how we can use the specifics of such animal studies to focus our attention, in the most productive manner, on making ourselves a more cooperative species.

Recall our guppy risk takers and what underlies their behavior. Evolutionary models predict this sort of cooperation when individuals have a high probability of bumping into each other and finding themselves in the same predicament again in the near future. Guppies show every sign that this is precisely what drives their risk taking, as they remember each other's actions for days and only cooperate with those that have reciprocated their kindness.

Can we learn anything about human cooperation from the guppy example per se? It depends. It would certainly be difficult to argue that any single example of animal cooperation is particularly useful to us in this context. Search hard enough in the animal world and you will find isolated examples of any behavior you can imagine, but isolated examples are hardly the framework on which to build a theory. Rather, it is the fact that underlying elements of the guppy example (multiple interactions among individuals, individual recognition) are played out over and over again in the animal world, in all sorts of species and in many different contexts, that makes the guppy case and those like it useful to us.

How might we go about using the moral of the guppy story to foster human cooperation? One way to address this problem is to construct a human scenario that probably has the same sorts of costs and benefits associated with it. What about guard duty in the army? Are not the costs and benefits of this situation similar to those in our guppy risk-taking scenario? Surely they are, in that one certainly receives a benefit if others on guard duty go out and inspect a potential danger, but if no one goes out, everyone is in trouble. It is important to stress here that I am comparing only the costs and benefits of risk taking in guppies and guard duty in soldiers. My example says nothing about how much more complex human soldiers are than guppy patrollers nor anything about the cognitive processes involved in each case; it simply looks at costs and benefits.

Given the analogies between the guppy case and the army case, what can we glean from them? When soldiers are placed in dangerous situations in which they must take turns assuming risks, the best social environment is one in which the military unit is relatively small and has been together for a long time, so that all members of the group know what to expect from each other and know that if they take risks, they will (or will not) be reciprocated. Memory, individual recognition, and scorekeeping all facilitate the use of the tit-for-tat strategy that we introduced earlier, and keeping fighting units small and stable should make such cooperation so much the more likely. This logic holds true for any human scenario that involves risks and turn taking, not just army guard duty. Always keep in mind, however, that in this example, like all that we will look at, we are not trying to copy (or not copy) nature, but rather to use what we have learned from nature to suggest where we should focus our moral compass.

Can we use the cooperating lionesses story to somehow foster human cooperation? Suppose that we wish to foster cooperation in a small community of people that is in need of many resources: is the lion example of any use? Female lions hunt cooperatively when going after large prey but hunt smaller morsels alone. In other words, when the resource is large enough to both require more than one hunter and be shared (if the hunt is a success), cooperation occurs; otherwise it does not. Is there some way to create an environment in our community of humans in which people need to go after large resource items that are divisible rather than smaller items that can be obtained by solo players? Can a community-based incentive system be created to facilitate this?

In many urban areas today, planting large-scale gardens to spruce up a previously neglected area is common. Such gardens could be managed in many different ways. One way would be to allocate a small amount of land to each participant, allowing him to reap the fruits he sows. A second plan would be similar but with a critical twist. In addition to being given her own plot of land to do with as she pleases, each gardener would need to agree to spend some small amount of time working on a communal section of the garden. Individuals would then get the fruits grown on their own area and some portion of the communal crops, but only if they cooperated in the communal part of the garden.

The guppy and lion cases are only two of the many animal examples that will be put forth to help us understand how to use animal studies to facilitate human cooperation. While I believe that my approach to these issues is novel and will shed new light on human sociality, it can be better understood once we have some history under our belts.

Darwin's Bulldog versus Russia's Anarchist

Given the monumental impact that Charles Darwin's works *The Origin of Species* and *The Descent of Man and Selection in Relation to Sex*[16] have had in both the social and the physical sciences, it is often surprising to many that Darwin's ideas with respect to natural selection are straightforward—in fact, remarkably straightforward. Consider any measurable characteristic of an organism—height, weight, ability to see, and so on. If variations in this characteristic exist (for example, differences in height among individuals in a population), and if there exists a means by which individuals produce offspring that resemble themselves with respect to this trait,[17] then any variant that outreproduces others will spread through the population over time. If taller individuals, perhaps because they have access to greater food resources, have more young, over time we expect to see the average height of individuals in that population increase. This argument holds true even if being just a little taller gives you a very slight edge in terms of the number of offspring you raise; through evolutionary time small differences can accumulate into large changes.

Behavioral biologists from the time of *The Origin* on have argued, as did Darwin himself, that the theory of natural selection applies not only to anatomical and physiological traits but to behavioral traits as well. In fact, one of Darwin's staunchest advocates was George Romanes, a founding father of social psychology.[18] The argument is the same as before: if a number of different *behavioral* options exist and there is some means for these behaviors to be transmitted across generations, then any behavior that has a slight advantage in terms of its effect on individual reproduction will increase in frequency. So too for cooperation: if it increases fitness (roughly speaking, number of offspring), it should increase in frequency in a population.

Darwin himself was quite interested in, and sometimes troubled by, the evolution of cooperative and altruistic behavior. Through his observations he concluded, for instance, that the "most common service in the higher animals is to warn one another of danger by means of the united senses of all."Yet not all of the cooperation in the animal kingdom was easily explained. Darwin viewed the self-sacrificial behavior of bees and wasps (stinging intruders to guard the nest, but usually dying in the process) with a degree of awe, but he also realized that such extreme altruism was a challenge to the very core of natural selection thinking. After all, how could giving up your own life in defense of your nest ever be selected? Darwin had the creative insight to come up with the solution, known as kin selection (although the mathematics were not actually formalized for another hundred years). Darwin recognized that cooperation plays an important role in man as well as in animals, and while he centered his argument on "primitive man," it takes little imagination to extend it to modern societies:

> When two tribes of primeval man, living in the same country came into competition, if (other circumstances being equal), the one

> tribe included a great number of courageous, sympathetic and faithful members, who were always ready to warn one another of danger, and to aid and defend each other, this tribe would succeed better and conquer the other. . . . It must not be forgotten that although a high standard of morality gives but a slight advantage to each individual man and his children over the other men of the same tribe, yet that an increase in the number of well endowed men and an advancement in the standard of morality will certainly give an immense advantage to one tribe over the other. (*The Descent of Man and Selection in Relation to Sex,* 1872)

It didn't take long for Darwin's ideas on natural selection applied to cooperation to stir up quite a hornet's nest. It was bad enough that Darwin's ideas shook up people's notion of their place in the universe, but now he was suggesting that our defining moral attributes—being compassionate and cooperative—are really just results of a history centered around chimps and other primates.

While Darwin believed that cooperation and competition were both prevalent in the animal world, two more extreme camps began to polarize the issue. On one side was none other than Darwin's friend, the eminent Thomas Henry Huxley, arguing that in the animal world cooperation was an anathema. Huxley, also known as "Darwin's Bulldog" because of his never-ending defense of Darwin's ideas, put forth his "gladiator" view of the world in the popular British newspaper called *The Guardian.* In Huxley's eyes the animal kingdom was a ruthless jungle, and a soft-bellied cooperator would stand no chance against a more cunning individual who would stop at nothing.[19] With respect to cooperation, Huxley had adopted Herbert Spencer's view of "nature, red in tooth and claw":

> From the point of view of the moralist, the animal world is on about the same level as the gladiator's show. The creatures are fairly well treated, and set to fight; whereby the strongest, the swiftest and the cunningest live to fight another day. The spectator has no need to turn his thumb down, as no quarter is given . . . the weakest and the stupidest went to the wall, while the toughest and the shrewdest, those who were best fitted to cope with their circumstances, but not the best in any other way, survived. Life was a continuous free fight, and beyond the limited and temporary relations of the family, the Hobbesian war of each against all was the normal state of existence. (*The Struggle for Existence and Its Bearing upon Man, 1888*)

This nasty picture—I certainly wouldn't want my kid going to school in that neighborhood—was vigorously opposed by many. In particular, the writings of Alfred Russell Wallace and Petr Kropotkin show a nature that is almost diametrically opposed to that painted by Huxley.

Alfred Russell Wallace had great insight but bad timing. He came up with the theory of natural selection at the same time as Darwin. Historians of science argue that Wallace was an even more avid "Darwinian" than Darwin himself in that Wallace believed in the almost unlimited, unending power of natural selection. Where Darwin and Wallace parted, however, was on the issue of humans, and particularly the human psyche. While Darwin viewed human evolution in the same light in which he viewed all evolution, Wallace did not.

Wallace felt that human morality and intelligence were somehow outside the realm of natural selection, and that "the whole reason, the only raison d'etre of the world . . . was the development of the human spirit in association with the human body." Although it is always difficult to prove such a claim, I'd venture to guess that Wallace's views were tied to his deep-seated religious beliefs about man's role in the cosmos, as Wallace essentially became a spiritualist and gave up science later in life. Given his views on morality and intellect, it should come as no surprise that he viewed cooperation as the norm in nature, rather than the exception. Contrast his remarks below with Huxley's gladiator view and you will see how radically two great minds can differ:

> On the whole, then, we conclude that the popular idea of the struggle for existence entailing misery and pain on the animal world is the very reverse of the truth. What it really brings about is the maximum of life and of enjoyment of life with the minimum of suffering. (*Darwinism*, 1891)

Prince Petr Kropotkin was an even more fascinating man than Wallace. Prince of Russia, Kropotkin gave up his royal position to become an anarchist (he was one of the founding fathers of the discipline), a geologist, and a natural historian. As a natural historian and traveler, Kropotkin was a keen observer of animal social behavior. Huxley's gladiator scenario could not be further removed from what Kropotkin saw on his journeys, and he engaged in a spirited written exchange with Huxley in *The Guardian*. This exchange led to Kropotkin's classic *Mutual Aid* (1908). In this wonderfully written book, Kropotkin tells of seeing animal cooperation (that is, mutual aid) at every turn: "In all these scenes of animal life which crossed before my eyes, I saw mutual aid and mutual support carried on to an extent which made me suspect in it a feature of the greatest importance for the maintenance of life, the preservation of each species and its further evolution."

Daniel Todes has recently put forth a fascinating hypothesis to explain the stark difference between the views of Kropotkin and Huxley.[20] It appears that many Russian naturalists and evolutionary biologists of Kropotkin's era believed that cooperation was the default state of nature, while most Europeans tended to line up more with the Huxley-like view that nature was much more nasty than cooperative. The difference, suggests Todes, is that Russian biologists and naturalists made their observations in rather hostile environments, such as Siberia, while most European biologists of the day were off studying in the tropics. So, not only were the Europeans getting a suntan, but they were observing an environment in which competition was the driving social agent. The Russians, however, were viewing animals that needed to cooperate with one another to fight against a harsh physical environment. We will return to this notion of cooperation in harsh environments later on.

Todes's arguments show once again how a behind-the-scenes view of science and scientists often reveals that unexpected, often idiosyncratic, factors can shape what is studied and why. Those researchers steeped in social behavior such as cooperation or even aggression may be more prone to having their ideologies affect their research. It is hard to imagine how your political leanings could affect your work if you were studying, for example, the shape of red blood cells in mammals, while it takes little imagination to see how this might happen if your research centered on the evolution of pro-social behaviors—cases in point being Wallace's spiritualism and Kropotkin's avid social libertarianism. Of course we all hold political views, myself included. No doubt my Judeo-Christian heritage, as well as my strong belief in one God who holds people accountable for their actions, affects every aspect of my life, in some form or another. How could such a belief system not? My hope, as I would assume is the case for most scientists who are also religious, is that while my belief system may draw me (directly or indirectly) toward certain scientific questions over others, it has no impact on my objectivity when performing experiments. *Something* is always responsible for drawing scientists to the questions they eventually address.[21]

The Kropotkin/Wallace versus Huxley debates on cooperation made for good reading and good natural history (two things that the British were quite fond of in the 1880s), but they did little to formalize an evolutionary model of cooperation that could help us predict when we should expect cooperation and when we should not, nor did they address the different types of cooperation seen in nature. These issues have been addressed only in the last twenty to thirty years. The Huxley/Kropotkin debates were mainly about how common cooperation was and why each side believed that they had painted an accurate picture of nature. To fully understand modern theories of cooperation, however, we need to grasp how the questions surrounding cooperation have been framed, and that is where history plays a role.

Prince Kropotkin was the most vociferous, but not the lone, spokesman for the view that cooperation is the norm in nature and that, left on their own, animals will naturally cooperate in many instances. Accordingly, if humans wish to be more cooperative, we must merely look to nature—which to Kropotkin and others of his day meant not only animals but primitive human societies—and imitate it. Again, Kropotkin in his own words: "If we resort to an indirect test and ask Nature: Who are the fittest: those that are continually at war with each other, or those who support one another? We at once see those animals which acquire mutual aid are undoubtedly the fittest" (*Mutual Aid*).

While it probably pulls on our heartstrings to believe that the natural world is quite a cooperative place indeed and that mimicking nature is the way to cooperation, the literature on every sort of noncooperative act imaginable suggests that this view is naive—nice in principle, wrong in fact.

On the flip side of the coin were those who believed that nature was a never-ending bloodbath and that little could be learned from observing the carnage. A more extreme position on this theme was put forward by Huxley, who essentially argued that rather than copying nature, we should be doing everything we can to oppose it. This was not just another in a line of questions Huxley addressed, but rather one of his passions. Huxley implored his audience to admit "once and for all, that the ethical progress of society depends, not on imitating the cosmic process, still less in running away from it, but in combating it."[22] But the looming presence of both cooperation and carnage in nature suggests that a simple guidepost like "do the opposite of what occurs in nature" is an argument that lacks much vigor.

Most evolutionary biologists reject both the "imitate nature" and "oppose nature" arguments and adopt what I will call "the one long argument" approach. In a wonderfully readable book entitled *One Long Argument,* Harvard University's Ernst Mayr, a founding father of modern evolutionary thinking, puts forth the idea that all of Darwin's numerous books and papers have one theme—that one must understand history and the process of natural selection, and its typical outcome of adaptive evolutionary change, to understand whatever one is studying.[23] While all reasonable biologists believe natural selection to be a powerful force, adherents of the one long argument approach argue that the facts of nature tell us nothing about morality. Sometimes natural selection favors cooperation, sometimes not. Stephen Jay Gould, among others, has popularized this general notion in his monthly articles in *Natural History* magazine:

> There are no shortcuts to morality. Nature is not intrinsically anything that can offer comfort or solace in human terms—if only because our species is such an insignificant latecomer in a world not constructed for us. So much the better. The answers to moral dilemmas are not lying out there, waiting to be discovered. They reside, like the kingdom of God, within us—the most difficult and inaccessible spot for any discovery.[24]

I take exception to Gould's characterization of humans as "insignificant latecomers" but agree that it is dangerous to use natural selection thinking to guide human behavior. Unfortunately, ever since E. O. Wilson published his classic book *Sociobiology,*[25] a number of biologists, psychologists, and anthropologists have used evolutionary approaches to make preposterous, unsubstantiated claims about human nature—on humans as inherently warlike (or inherently peaceful), on the nature of sexuality and parenting, and so on. This is so common that it even has its own name—pop sociobiology—and it has rightly been criticized by many. The danger of this sort of thinking, as Robert Wright points out in *The Moral Animal,*[26] is that it leads one to the mistaken impression that feelings like sympathy, guilt, and the notion that right should be rewarded and evil punished are really just natural selection manifesting itself *and should circumstances change, natural selec-*

tion may simply favor the converse emotions in humans. This is an unacceptable thought for anyone with a sense of absolute right and wrong.[27]

The scientific brute facts of nature tell us nothing about what is a moral act and what is an immoral act—they could not possibly do so, for that is not what science is about. Whether a lion killing a gazelle is a moral act is not something science can address, as there is no experiment that would allow one to come to that conclusion or to reject that proposition. I do nonetheless believe that studying animal examples of cooperation (or the lack of it) can be quite useful in helping us structure human interactions.

Animal cooperation often shows us, as I suggested earlier, what to expect when the complex web of human social networks, as well as the laws and norms found in all human societies, are absent, and acts as a sort of baseline from which to operate. I am not arguing that animal cooperation is the rule or the exception in nature, just that it occurs often enough to make it an irresistible subject. Moreover, while natural selection is the driving force shaping cooperation in animals, and I present animal examples as a means to enhance human cooperation, I am not arguing that natural selection thinking should be the guidepost we use in shaping our own behavior. And I am by no means falling into the trap of the "naturalistic fallacy"—that because something "is" in nature, then it "ought" to be that way.[28] Whether natural selection favors something in nonhumans, or even whether natural selection might favor a behavior in humans if we stripped away culture, is not the primary issue with respect to shaping human cooperative tendencies. Rather, the critical point is that we can use animal examples as a means for focusing our moral compasses on the right elements, from the right perspective, in such a manner as to make human cooperation more likely.

The logic underlying the guppy and lioness examples alluded to earlier, and how such cases can be used to foster human cooperation, form the framework on which this book is built. We will walk our way through each of the four paths to cooperation, looking at how and why they work and the controversies surrounding each. Investigating often dazzling, almost incredible examples of cooperation in nonhumans will take us to all corners of the earth.

1 All in the Family

And the man knew his wife Eve; and she conceived and bore Cain . . . and again she bore his brother Abel. . . . And in the process of time it came to pass, that Cain brought of the fruit of the ground an offering to the Lord. And Abel, he also brought of the firstling of his flock and of the fat thereof. And the Lord had respect unto Abel and his offering, but unto Cain He had no respect. And Cain was very wroth, and his countenance fell. And the Lord said unto Cain: "Why art thou wroth? and why is thy countenance fallen? If thou doest well, shall it not be lifted?"

—Genesis 4: 1–7

According to the Old Testament, the first human siblings were not particularly fond of one another. Fair enough—we all know of cases of sibling rivalry, not to mention the animosity that we might hold for some of our more distant relatives who pop in for the occasional holiday dinner. But, in general, blood really is thicker than water despite tales of murderous brothers. Later in the story of Cain and Abel we encounter a poignant question: "And the Lord said unto Cain: where is Abel thy brother? And he said 'I know not; am I my brother's keeper?' " The answer, at least in terms of how we behave, is a qualified yes—you are indeed your brother's keeper.[1]

If your brother and a stranger were drowning, who would you save first? The reply most of us would expect—"my brother, *of course*"—suggests, but certainly does not demonstrate, the importance of the role of kinship in structuring human cooperative acts. Is it possible, however, that the reason we might choose our brother has nothing to do with the fact that he is a blood relative per se? Is it merely the fact that we have spent so much time with our siblings that drives our actions? A simple thought experiment might help us to understand whether this true. Imagine it is not a stranger with your brother there in the water, but your closest friend. How would you feel in that case? How would most people react? If kinship was the overarching theme, most should still reply "my brother, of course." This train of thought has led behavioral ecologists to appreciate just how important kinship is in the human social dynamic.

Family Accounting Schemes

The scenario in which you save your brother or someone else functions as a nice illustration of how kinship affects human cooperation, but it is not something one envisions as a starting place for significant scientific breakthroughs. Or is it? Evolutionary biologists' first introduction to the notion that blood relations affect social behavior actually came in the 1930s in a form similar to this example. It was then that J. B. S. Haldane, a founder of modern evolutionary theory, suggested that he would risk his life to save two (but not one) of his brothers and eight (but not seven) of his cousins. Haldane, quite versed in mathematics, made this rather bold statement by counting copies of a gene that might code for cooperative behavior. Such a gene-counting approach to kinship and the evolution of cooperation has been extended by theoreticians but in its most elementary form is the heart and soul of kinship theory.

Let us see how this idea works and how it has been formalized into what is known as kin selection or inclusive fitness theory.

The evolutionary biologist's definition of relatedness and kinship may strike many as surprising, if not odd. In such a definition, relatedness centers on the *probability* that individuals share genes that they have inherited from some common ancestor (parents, grandparents, etc.). A jargon phrase summing up this approach in behavioral ecology is "identity by descent." For example, you and your sister are kin *because* you share some (in this case many) of the same genes and these have been inherited from common ancestors, mom and dad. Similarly, you and your cousins are kin, because you share genes in common (not as many as siblings) and common ancestors, your grandparents. Common ancestors are the most recent individuals through which two (or more) individuals can trace genes that they share in common.

Once we know how to find the common ancestry of two or more individuals, we can calculate their relatedness, which simply amounts to the probability that they share genes that are identical by descent—genes that have been inherited from a common ancestor. In the literature on kinship, this probability is often labeled r (for "relatedness"). For example, you and your brother are related to one another by an r value of 1/2.[2]

From a "gene's-eye" perspective, calculating relatedness is the first critical step in understanding how kinship can favor cooperative behavior among individuals. Genes' survival depends on the number of copies of themselves that they get into the next generation. This is often thought of in terms of what effect a given gene has on the individual in which it resides, but relatedness suggests that this is a myopic view. If relatives have a high probability of sharing a given gene, then that gene can potentially increase its chances of getting more copies of itself into the next generation by coding for some behavior that helps relatives. Again taking a gene's-eye view, relatives are just vehicles who are likely to have copies of you (the gene in question) inside them as well.[3] But, and this is a big "but," relatives only have some probability (*r*) of having a copy of, for example, a gene for cooperation. A gene in sibling 1 "knows" that a copy of itself may reside in sibling 2, but only with a 50 percent probability. The more distant the relative, the less likely a copy of the gene resides in them as well. So, phrased in the cold language of natural selection, relatives are worth helping in direct proportion to their relatedness. This is because relatedness is a measure of genetic similarity, and genes are the currency of natural selection.

Behavioral ecologists are not so foolish as to assume that animals are able to calculate relatedness in the manner described above. We only assume that natural selection favors individuals who act in ways that make it appear as though they are able to make such calculations. How animals determine who is kin and who isn't is a matter of some debate these days.[4] For example, one theory suggests that animals determine relatedness by matching a suite of traits (a template) that they possess against the same suite of traits in another individual. Depending on the degree to which traits match up, individuals are treated as full siblings (if many matches occur), half siblings (if fewer matches occur), cousins, and so on, down to the category "unrelated individual" (if, for example, no matches occur). Such "matching games" have their flaws; mistakes can be made in determining the level of overlap, and some relatives may erroneously be treated as nonrelatives, while some nonrelatives may be viewed as relatives.[5] Often, however, rather than a suite of traits, a single characteristic is used to determine whether another individual is kin and if so, what type of kin. In many insect species, for example, kinship is assessed by odor. Individuals who smell like you (or your nest) are relatives, and how closely related they are is determined by how similar their odors are to yours.

While most behavioral ecologists accept that such matching (either of many cues or a single cue) is important, they believe that there is another, simpler explanation for how animals determine who qualifies as kin, an explanation that I'll refer to as the "no place like home" hypothesis. Under this hypothesis, animals simply treat all others that grew up in their nest (territory, burrow, etc.) as relatives. This very simple rule is often quite powerful. With the exception of some species that try to trick other species into raising their offspring, the odds are quite strong that those who grew up in your nest are in fact your siblings and parents.

The details of how animals evaluate relatedness are fascinating, but all we really need to know to examine kin-selected cooperation is that many animals do in fact behave in ways that allow them to distinguish between kin and nonkin and even to distinguish between different degrees of relatedness. Once we have calculated relatedness, we are very close to reaching a general rule for when cooperation among relatives should be favored and when it should not. We need only consider two more factors: the cost of the action to the individual cooperating and the benefit to the recipient of such a cooperative act. Let us call the cost of a cooperative act to the donor c, and the benefit to the recipient b. In 1964, W. D. Hamilton (now at Oxford University) showed that cooperation among relatives should evolve when the following holds true: $r \times b \geq c$.

In other words, cooperation among relatives is favored if, and only if, the benefit of the act multiplied by the relatedness of the actors is greater than or equal to the costs (≥ is read "is greater than or equal to"). This equation, $r \times b \geq c$, has become known as Hamilton's Rule. Essentially, Hamilton's Rule says the following: There is some cost (c) that "must be made up for" if the gene for cooperation is to evolve, as cooperating with others is often a risky business. One way to make up for this cost is through the benefits (b) a relative receives, because relatives may carry the gene for cooperation as well. But, relatives have only *some probability* of carrying the cooperation gene and so the benefits received must be devalued by that probability. If I pay a cost for undertaking an action, but there is only a probability that I will receive indirect benefits (in this case through my relatives), I need to factor that into my equation and that is just what r does.

We can illustrate the use of relatedness to predict cooperation among kin with a simple chart. Consider an action that you take that reduces your chances of survival by 50 percent (a very serious cost) but increases the probability of survival

IS SCREAMING AT THUGS WORTH IT?

Relative	Degree of relatedness	Number of relatives who must hear your screams
Sibling	1/2	≥2
Parent	1/2	≥2
Grandparent	1/4	≥4
Grandchild	1/4	≥4
Uncle/aunt	1/4	≥4
Cousin	1/8	≥8
Spouse	0.0	—

of the relative(s) you are trying to save by 50 percent *each* (a considerable benefit). Such extreme costs and benefits might, for example, mimic a situation in which you scream out when a gang of armed thugs is approaching. This serves to announce the presence of thugs to the relatives around you, but at the same time the scream draws the marauders' attention your way, a dangerous action indeed. Based solely on kin selection theory, the table shown here outlines the number of relatives that need to hear your scream before natural selection alone would favor such dangerous behavior on your part.

The table illustrates the fundamental point of inclusive fitness theory: the greater the degree of relatedness between individuals, the more likely that kin-selected cooperation is selected. There need only be two (or more) siblings around for you to make that scream, but you'd need eight or more cousins present (a much less likely event), if they were the only relatives in the vicinity! How exactly, though, do we use Hamilton's Rule to come up with the correct number of relatives in the table? Consider the case for siblings. If a single sibling hears an alarm call, then $r = 1/2$ and b and c are still each $1/2$. In that case r multiplied by b is not greater than c, Hamilton's Rule is not met, and cooperation via kinship is not favored by natural selection.

Suppose, however, that three siblings hear the alarm call. Now b is tripled (three recipients), but c is the same (the alarm call still draws the predator's attention), so $r \times b = 1/2 \times (1/2 \times 3)$ for a total of $3/4$, which is greater than c, and Hamilton's Rule is satisfied. The same logic can be applied to any relative in the table (or for that matter, any relative not in this table). Take note, as well, that the relatedness of an individual to his/her spouse is 0 (with the exception of marriages among relatives).

Although one's spouse is kin in the everyday usage of the term, we don't generally share genes inherited from a common ancestor with our spouses and hence this category of relative is in effect removed from kin selection theory.

Of the four paths to cooperation that we will focus on, kinship is the best understood, most accepted, and least controversial. It is in every legitimate textbook on evolution and is cited in more papers in the field than any other set of theories. There is even a belief among some evolutionary and behavioral biologists that Hamilton's work in this area marks the start of the modern discipline of behavioral ecology. But even kin selection theory is not without its controversies.

One area of contention with respect to kinship and cooperation centers on whether it really matters where the genes we are counting are located.[6] Kin selectionists correctly argue that blood relatives are more likely to carry the same gene than are individuals drawn at random from a population. But what if some other mechanism besides kinship could create groups in which individuals were all likely to carry one or more genes coding for cooperation? Does it really matter that such individuals don't share other genes, like kin do? After all, we are interested in the gene(s) coding for cooperation, and everything else is in some respects background material for that gene. Who cares whether individuals carry the same genes because of kinship or for some other reason—shouldn't the process by which cooperation is selected for work just as well in both cases? The answer to this question, as Hamilton himself noted, is yes, the process works the same; whether individuals share the gene(s) for cooperation because of relatedness or some other factors is irrelevant.[7]

Yet kin selection advocates are not so fast to roll over. Sure, mathematically speaking, you are right, they say, but in practice the distinction we are arguing about is still real and important. Give us, they say, a good example of how individuals sharing a gene are brought together, if relatedness (which automatically brings them together) is not in force. The answer typically given by kin selection critics is that individuals that share a gene for cooperation may gather together specifically to be near other cooperators, because cooperators do particularly well when around others like themselves and so should choose this option, when it becomes available. "Be specific," say kin selectionists, "give us a *real* example." And this is where the kin selectionists start looking a bit better than they did after losing the mathematics argument, because behavioral ecologists are usually stopped in their tracks when it comes to finding a good animal example to answer this question.

Although such examples may be hard to uncover in animals, those interested in enhancing human cooperation argue that the evidence for cooperators choosing other cooperators as partners in our own species is anything but scarce—even when kinship is not in play. How others will act is one *primary* means by which we choose with whom we will interact. So, for humans then, while kinship is an extremely important force selecting for cooperation, there are many other ways cooperators may cluster together aside from kinship. We need to recognize this in our behavioral studies and our conjectures about human cooperation.

The above controversy is admittedly a semantic one in part, but semantic arguments can be quite illuminating. Hamilton's Rule—which in words roughly translates to "all else equal, cooperation should be most common among close relatives"—is as close as behavioral ecologists get to a "law of nature." It is an underpinning of all modern evolutionary approaches to social behavior and is, in many ways, as much an approach to behavioral biology as it is a theory. The data gathered to date certainly support the claim that Hamilton's Rule is extremely powerful. It is not a "law" in the sense that gravity is, but it is about as near to one as behavioral biologists can hope to come, given the astonishing complexity and variability that is an inherent part of the subject matter they tackle.

From the standpoint of reputation, Hamilton's Rule was quite good for the field of behavioral ecology, at least in one sense. While solid mathematical theory has been part of evolution since the seminal work of J. B. S. Haldane, Ronald Fisher, and Sewall Wright in the 1930s,[8] it was not truly a centerpiece of evolutionary approaches to behavior until Hamilton's Rule. For many in the field of behavior, there was an unspoken envy of the hard sciences (physics, chemistry, even other parts of biology) that had steadfast "rules" that could be written out for skeptics (not to mention funding agencies). Hamilton's Rule provided such ammunition to behavioral ecologists. Let's take a look at some examples of why this is so, with a few cases from the animal kingdom, before moving on to how such scenarios can help us foster human sociality.

The Insect Police

The so-called social insects have been a godsend for advocates of kin-selected cooperation. The reason lies, at least in part, with the bizarre genetics of social insects such as bees, wasps, and ants (collectively known as hymenopteran insects). Humans (and most other animals) are diploid organisms, which means that we have two copies of each of our chromosomes. Our forty-six chromosomes are twenty-three matched pairs. The only stages of human life that are not diploid are sperm and egg, as they have only a single copy of each of our twenty-three distinctive chromosomes. Sperm and egg then are called haploid rather than diploid. Of course, sperm and egg later fuse to form diploid animals.

Much of life on earth, such as bacteria and viruses, is always in the haploid phase. Why some life on earth is diploid and some haploid is a fascinating question, but not one critical to the issues we are examining. What makes bees, wasps, and ants so bizarre is that females are diploid and males are haploid—a genetic system known as haplodiploidy. What this means is that when a male fertilizes a female, only daughters are produced because the sperm and egg fuse to produce a diploid creature, and in most social insect species diploids are female. Females, however, produce sons from unfertilized eggs (eggs that have not fused with sperm)—which means that sons never have fathers!

Haplodiploidy creates some very strange scenarios. In diploid and haploid creatures, relatedness between two individuals is symmetric; that is, if a father is related to his daughter by an *r* of 1/2, then a daughter is related to her father by the same value. Not true for the social insects. To see why, focus your attention on the father/daughter relationship. Fathers are haploid and give a copy of each chromosome they have to their daughters. Hence fathers are related to daughters by a value of 1. Daughters, however, are diploid, in that they get one copy of each chromosome from each parent, both mom and dad; so a daughter's relatedness to her father is 1/2 (half her chromosomes come from dad)—fully half of her father's relatedness to her.

The most relevant effect of the strange genetics of the social insects is its impact on average relatedness within insect colonies. Before seeing this in detail, keep in mind that in many social insect colonies a single queen produces all the offspring for a group. This means that the vast majority of individuals in such colonies are sisters and brothers. Haplodiploidy has the twofold effect of making sisters "super-relatives" and making the relatedness between brothers and sisters only half of what it is in diploid brothers and sisters. Sisters end up with a relatedness value of 3/4 (50 percent greater than the same relationship in diploid species), and sisters are related to brothers by a value of 1/4 (half the value found in diploid creatures). So, a clear prediction from kin selection theory is that since females are much more related to their fellow colony members than are males, when colonies have more females than males (as in most social insects), cooperation should occur predominantly in this sex. And of course it does, as "workers" in insect colonies are almost always female! It is females that sacrifice their lives by stinging folks and ruining an otherwise pleasant summer day. It is also females that undertake virtually all of the everyday activities that keep a colony functioning—food gathering, care for the young, and so on. One particularly interesting and unique behavior found among female social insects is "policing" behavior.

Bee colonies are rightly thought of as models of both efficiency and harmony. It is mind-boggling what a colony of tiny insects can accomplish in a short period of time: regulating the temperature of a hive, caring for young, defending against many predators, finding food, recruiting others to join in bringing back the booty, and a myriad of other activities. Some of this efficiency (and harmony) has been attributed to a single queen often producing all of the eggs for a colony, thus allowing worker females to spend their time on other hive-related necessities. Queens accomplish this enviable task by using a barrage of chemicals to inhibit other females—the workers—from reproducing. Yet, as with any chemical inhibition system, it is inevitable that some workers will escape these anti-aphrodisiacs and thus will have a much greater chance of reproducing than their subdued sisters. Once this fascinating new door is opened, we can ask whether kinship theory can guide us with respect to a rather nasty question: should the eggs laid by workers that ignore the queen's chemical castration cues be left alone by their sisters or vigorously attacked? The answer to this question is rather personal, if you happen to be the queen, as it depends on how many males you opt to mate with.

The relatedness between individuals in an insect colony depends on how many males inseminate the queen.[9] The more males the female mates with, the more different lineages there are in a colony—each line's ancestry going through the queen and a given male. Once again, however, the strange genetics of such insects creates a novel situation. Rather than showing a family tree more complicated than that of the British monarchy, it can be shown that *if* the queen of a colony mates with a single male, then female workers in the colony turn out to be more related to nephews than to brothers. If the queen mates with numerous males, however, that situation reverses itself and female workers in a hive are now more related to brothers than to nephews. We shall focus on the second scenario because of the fascinating kin-selected cooperation emerging from it.

Whether females in a social insect colony are more related to brothers than to nephews can have quite serious implications about when and whether we should see kin-selected cooperation, and if it exists, what form such cooperation should take. To see this, first recall that brothers are those individuals produced by the queen, while nephews are those produced by sisters that have somehow managed to avoid the queen's chemical anti-reproduction agent. A conflict of interest then arises between sisters that have managed to escape and those that have not.

Aside from the queen, females who can reproduce (i.e., those that do not fall victim to the queen's attempt to monopolize reproduction) are always selected to do so. When females reproduce they always produce males, since such females are almost never inseminated. This creates a problem, however, for those females who can't reproduce, as they are more closely related to the queen's offspring (their brothers) than to their sisters' children (their nephews). Kin-selected cooperation on the part of those nonreproducing female workers then favors any action that increases the odds of the queen's offspring surviving at the cost of nephews.

There is little a female can do to stop one of her sisters from reproducing, if her sister has avoided the queen's attempt to do so already. But there are options available. Once a worker has laid eggs behind the queen's back, her sisters could, for example, refuse to care for and help nephews. Or they could take more drastic action—they could eat eggs destined to be their nephews! Francis Ratnieks and Paul Visscher examined this possibility in honeybees, where females mate with ten to twenty different males.[10] Their results were astonishing. Those honeybee females who did not produce offspring "policed" the reproductive actions of their sisters.[11] If their sisters produced eggs on the sly, policing females destroyed the eggs. Ratnieks and Visscher found that honeybee workers showed remarkable acumen in discriminating between sisters' eggs and the queen's eggs. In a controlled laboratory setting, after twenty-four hours, only 2 percent of the sister-laid eggs remained intact, while 61 percent of the queen-laid eggs remained unharmed! But, given that the actual act of egg laying is rarely observed, how could honeybees know which eggs were laid by sisters and which by the queen? The answer appears to be that eggs are chemically "marked," such that queen-laid eggs smell different from worker-laid eggs. Why eggs should be marked so is still unclear, but one tantalizing possibility is that the queen marks her eggs to encourage workers to police the activities of their sisters.

Kin-selected policing is qualitatively different from the other types of cooperation so often found in animals. Rather than having individuals form a cooperative unit to accomplish some task, cooperation in honeybee police work takes the form of stopping others from cheating—a more subtle and complex action. At a more fundamental level, policing is powerful,[12] because it provides a direct deterrent to cheating, whereas in many other cases, we simply rely on cooperation being somehow more profitable than cheating, and this holy grail is often difficult to obtain.

There are many other cases of cooperation in highly related social insects.[13] I'll mention one other curious example: honeypot ants. In one species of these ants, the largest individuals actually hang from the top of a colony and act as living storage tanks for water and sugar. These "honeypot" individuals have soft and elastic abdomens, and if you watch long enough you will see other individuals come up and "turn on the faucet" to drink the resources stored there.[14] For significant periods of time, honeypot individuals do nothing but hang from the rafters and supply this service.

It is fascinating to find policewomen and living storage bins in the insect world, but how much of the cooperation we see is strictly due to the bizarre haplodiploid genetics of social insects? Can we expect anything so dramatic among mammals?

"Eureka!" Naked Mole-Rats

Physics is not the only discipline in science that has "Eureka!" stories. Just as physicists can recount the bathtub adventures of Archimedes and his famous exclamation when coming up with his theory of buoyancy (specific gravity), so too can the ardent student of behavioral ecology recite the story of Richard Alexander and Jenny Jarvis's discovery of extraordinary cooperation in a bizarre creature, the naked mole-rat.[15] Alexander, a professor of biology at the University of Michigan, traveled to various universities in the 1970s, giving lectures on the evolution of social behavior, particularly cooperative and altruistic behavior. One of his themes was why, despite significant effort, extreme sociality (like that seen in insects) had not been uncovered in mammals. Alexander described the characteristics he believed a mammalian system would need for insect-like ultrasociality to exist. He outlined a hypothetical creature that would undertake altruistic acts for relatives who lived in a safe environment with lots of food. He went so far as to give details: the species would eat large tubers (potato-like foods) and live in burrows in a tropical spot that had clay soil.

One day in May 1976, Alexander presented these ideas to some folks at Northern Arizona University. Afterwards, he was approached by someone in the audience who told him he had given a perfect description of the naked mole-rat of Africa. On the advice of this fellow, Alexander contacted Jennifer Jarvis (at the University of Cape Town), who knew more about naked mole-rats than anyone in the world. After much back and forth, which included trips by Alexander and his colleague Paul Sherman to Africa to actually see the creatures, Jarvis and Alexander realized that they indeed had found the first eusocial (ultrasocial) mammal.

After all the attention this animal has attracted from both scientists and the media, it is almost disappointing to see how bland the native habitat of the naked mole-rat actually is and how ugly these creatures are, even by rodent standards! Naked mole-rats are hairless and blind, with crinkled skin and two large incisor teeth sticking out from their mouths. And those are the adults; the babies are even harder to look at for very long. First collected in Ethiopia in 1842, naked mole-rats (whose scientific name is *Heterocephalus glaber*) live within groups averaging about seventy individuals (but ranging up to almost three hundred) in underground burrows, from which they rarely, if ever, emerge. Such burrows average about two miles in length.[16] Naked mole-rats have been studied primarily in Kenya and are often found in arid areas covered with dust and brush. Typically found near dirt roads, colonies can be located by molehills that pock the landscape. But what naked mole-rats lack in beauty and scenic living conditions, they make up for in fantastic behaviors.

One female alone (among many in the colony) is responsible for all the reproduction in a naked mole-rat group (three or so males in the group are responsible for the male side of mating). No other mammal that we know of, except another species of naked mole-rats discovered later on, has a single "queen," and this finding sent shock waves through the behavioral biology community. Kin selection theory suggests that such extreme cooperation, wherein most individuals give up the opportunity to reproduce, should be limited to species in which individuals are somehow extremely related to each other, yet naked mole-rats are mammals and don't have the bizarre genetics that allow for the "super-relatives" we saw in the bees and ants. So how could such a bizarre system have evolved here? Before answering this question, let's get a more comprehensive sense of just how much cooperation goes on among these creatures.

The queen and the handful of males she mates with have a twofold advantage over others in naked mole-rat colonies: not only do they monopolize all colony reproduction, but they also live longer than their nonreproducing colony-mates. Yet in the relatively short time that nonreproductive males and females are around, they get a lot accomplished, and without their cooperation naked mole-rat colonies would surely come to a screeching halt. In fact, those individuals not specialized in reproducing take on virtually all of the everyday cooperative actions that are the very lifeblood of colony existence. They excavate new tunnels (an absolutely critical aspect of colony survival), sweep debris, groom one another as well as the queen, and take on the unenviable and dangerous task of defense against predators.[17]

Why such dramatic examples of cooperation in a single species? What singles out naked mole-rats? The answer probably lies in kinship within colonies. As we mentioned earlier, naked mole-rats do not have the strange genetics of some insects, but they have managed to achieve the highest average relatedness on record for naturally occurring mammals. DNA fingerprinting (the same technique we read about in criminal cases) showed that the average degree of relatedness among individual naked mole-rats in a colony was a whopping 0.81 (out of a possible score of 1).[18] To put this in some perspective, unrelated individuals have a value of 0.0 for this indicator, brothers score (on average) 0.5, and the most related of all individuals, identical twins, score 1.0. So naked mole-rat individuals on average fall between ordinary siblings and identical twins on a relatedness scale, and even lean toward the identical twins' side of the equation. What a wonderful finding in support of kin selection and cooperation! Question: Where do we find the highest recorded degree of cooperation among all mammals? Answer: Just where kin theory says we should—in a species with uniquely high degrees of relatedness among group members.

Cooperation in the naked mole-rat even exceeds the borders of a single colony. As in many species that live underground, founding a new colony, which entails coming above the surface, is a very dangerous activity. In addition to being away from the food source and potential mates, the brave mole-rat who tries to start a new colony may find numerous predators lurking above ground. In fact, for quite some time it was believed that new colonies formed when larger established colonies split in two, thus alleviating the problems associated with a single individual surfacing to hunt for a suitable place to start a new group. It turns out that this picture is not accurate, at least some of the time. M. J. O'Riain and his colleagues found that colonies did not undergo fission to form new groups, but that specialized individuals took on the dangerous task of colony founding, thus making it unnecessary for their kin to risk life and limb themselves.[19] Such "dispersing cooperators" put on weight very quickly during development and were considerably larger and bulkier than other colony members, presumably making them fitter for the trials and tribulations of colony founding in a hostile environment.

Before labeling naked mole-rats as the perfect example of kin-selected cooperation, we must deal with one potential fly in the ointment. One might think that in such a cooperative system, the nonreproductives would *voluntarily* yield reproduction. This is, apparently, not quite true for naked mole-rats. Nonreproductives are coerced into yielding the act of producing offspring by aggression on the part of the queen; that is, reproduction by the masses is suppressed, not freely handed over to a single individual.[20] Suppression, however, may not pose as great a problem to cooperation in this system as it seems. We would expect there to be tremendous selection pressure on the nonreproductives to breed if it would increase their fitness, suggesting that direct reproduction is not the best route to increases in fitness for such individuals. In essence, what appears to be suppression on the part of the queen may be the only mechanism available to increase the fitness of both queen and nonqueen, making such apparently suppressive acts really cooperation on the part of all. This remains to be seen, but the suppression = cooperation argument certainly is an intriguing possibility.

Even aside from its Eureka-like discovery, naked mole-rat ultracooperation is truly astounding. No other mammal species has a single queen producing offspring that undertake such a wide variety of cooperative and altruistic endeavors. But then, no other mammals are as closely related to one another either.

In-House Baby-Sitters

How many of us wish we could convince our older child that staying home to watch a sibling is a more noble and rewarding act than going to a friend's party down the street? Finding a person to watch your child while both parents work is an even greater dilemma. Dwarf mongooses, however, seem somehow to have solved this very problem in a rather cooperative fashion, based primarily on kin bonds.

Dwarf mongooses are small, social carnivores that typically occupy dens in the savanna habitat of Tanzania, in groups of approximately ten to twenty individuals. A pack of mongooses usually consists of a breeding male, a breeding female, and their young from a number of consecutive broods (i.e., young of various ages) as well as an occasional immigrant from elsewhere.[21] Young "helpers" cooperate with their kin in a wide variety of activities, including feeding, nest defense, grooming, and transporting the young. But the most fascinating and interesting variant of kin-based cooperation may very well be "baby-sitting."[22]

A typical day for a mongoose pack in the Serengeti begins with the adult male and female, as well as some of the immature adults, leaving the den to search for food. But what of the almost ever-present very young mongooses? Are they left alone at the burrow to take care of themselves, undefended? Not according to Jon Rood, who conducted field studies on dwarf mongooses for many years. Rood found that out of eighty-five observations recorded, in all but a single case there was at least one baby-sitter at the nest. Baby-sitting, almost always undertaken by kin (older siblings), provided a critical service as very young individuals left alone in the den are particularly susceptible to predation. Baby-sitters have been seen giving alarm calls and even chasing potential predators from the den. Cooperative kin then forsake foraging themselves to help watch over and defend their siblings.

Without the quintessential refrigerator to raid, what benefits are baby-sitters getting for their services? Clearly, helping and cooperating are primarily driven by kinship in this example, as most (if not all) young in a den are related. Save your sibling from being eaten and you save potentially many copies of genes that also reside in you (keep in mind again that this is how kinship is defined in evolutionary biology). Family ties, however, do not explain the whole picture. For example, Rood found that one pack he observed contained an immigrant, unrelated two-year-old female he named Carrie. Carrie not only undertook baby-sitting services, she was the group's predominant dispenser of child care. Rood suggests that one possible reason that unrelated individuals may baby-sit is that this act increases survivorship of all group members. The recipients of help today, even if unrelated, may be the alarm callers of tomorrow, benefiting the baby-sitter, albeit down the road a bit.[23]

Still, baby-sitters are temporary and not around all that much. Is anything more permanent in animal child care services available?

Cooperative Breeding in Bee-Eaters

In 1935 Alexander Skutch, naturalist and ornithologist extraordinaire, coined the term "helpers-at-the-nest" to describe a strange phenomenon he had observed in a number of species of birds. Skutch found that younger individuals, who were physiologically capable of reproducing, were not leaving their natal nests to find a territory and a mate. It was surprising indeed to find young individuals that could breed not cashing in on the opportunity. More fascinating, however, was the observation that such individuals were actually helping to raise the batch of babies (their new brothers and sisters) born during recent breeding cycles. Such helping often lasted significant periods of time and involved a suite of activities that included, but were not limited to, feeding the young and defending the nest against attack from predators.[24]

Some sixty years after Skutch introduced the notion of helpers-at-the-nest, there is still controversy surrounding the question of why capable young birds remain at home and help raise their baby brothers and sisters. One possible solution was put forth by Jerram Brown, who argued that the evolution of helpers-at-the-nest is really a two-part story.[25] He hypothesized that in many species, it is often quite difficult for young individuals to find suitable areas for starting their own new nests. Either there are no available slots in their environment or the usable area is sub-par, perhaps having too little food or too many predators. This situation creates pressure to remain at the home nest, past the age when individuals are physically able to breed on their own. Once the decision to remain home is in place, then such individuals help because it increases their inclusive fitness to do so (the critical comparison being with staying at home and not helping). So helpers do not stay at their natal nest because the inclusive fitness of doing so is higher than it would be on a good, new territory. Rather, they are forced by ecological factors to remain where they are, and as long as that is the case, the most productive activity for them to be involved in is raising their kin.

Lake Nakuru National Park, Kenya, is home to one of the better-studied species of birds with helpers, the white-fronted bee-eater. Stephen Emlen and Peter Wrege have been studying color-banded bee-eaters in various populations throughout this park for the last twenty-odd years, and they have come up with some remarkable findings.[26] Bee-eaters are a particularly nice species in which to study the phenomenon of helpers-at-the-nest, because their ecology and population structure allow us to investigate two components of kin-selected cooperation that are often difficult to study: (1) whether individuals, when given the option, help those they are most related to, and (2) the effect of helpers on the survival of their younger siblings.

Bee-eaters exist in extended family groups, wherein a number of breeding pairs of adults produce offspring during a given season. Such "clans" typically number three to seventeen individuals, including up to five breeding pairs and many unpaired younger individuals. As such, potential helpers can choose not only whether to help but whom to help. Possible recipients range from siblings to more distant relatives to unrelated young. As predicted by kin selection theory, helpers consistently choose whom they will help based on relatedness. In the 108 cases measured by Emlen and Wrege, helpers provided assistance to the *most* related individuals in their nest 94 percent of the time.[27] Helpers, then, are obviously helping the right individuals (according to kin selection theory at least), but do their actions really matter?

One initial critique of work on helpers came from researchers who did not think it mattered very much whether helpers were present or not. Whether helpers stick around the nest and *appear* to be helping their siblings is not the issue, the critical factor is whether one can *demonstrate* that they are helping—that is, increasing the number of young that survive.[28] Bee-eaters may prove to be an extreme case on the continuum of how helpful helpers truly can be, because the effects of helpers in this species are dramatic. The average productivity of nests without helpers is doubled with the addition of each new helper.[29]

Helpers have such a dramatic effect on the survival of their siblings, in fact, that parents have developed unique strategies to keep them around. By being particularly good at raising their brothers and sisters, helpers have made themselves a very valuable commodity, so valuable that parents would rather a helper remain at the nest than even attempt to breed elsewhere.[30] As a result, older bee-eaters (almost always fathers) actively interfere with their sons' attempts at breeding somewhere else—surely not what one would initially expect if kin selection was a large player in this system. That is, one might think that kin selection theory would often favor having parents do whatever they can to increase the number of grandchildren they have. Grandchildren, after all, are kin as well (albeit not as close kin as offspring). But a father actually increases his inclusive fitness more if a helper stays home than if it leaves. That is, from the perspective of the parent, the number of copies of its genes making it to the next generation would be higher if helpers stayed at the nest and helped raise their sibs (the parents' offspring) than if they attempted to reproduce on their own (and produce grandchildren). Since dad is in control, in the sense of being larger and more powerful, he calls the shots, and kin selection favors disrupting junior's attempts at leaving home. The end result is that kin-selected cooperation will *usually* make it worthwhile for a helper to stay, but even if that is not the case, kin-selected aggression will see to it that he stays, regardless.

Guideposts

The importance of kinship in human social dynamics is so much part and parcel of our everyday lives that it has worked its way into popular literature. In *Slapstick: or Lonesome, No More!* Kurt Vonnegut suggests that to make the world a better place we need to create a series of ever-extending artificial families.[31] In such a world, you could tell if someone were in your artificial family simply by their name. Names in Vonnegut's fictional land were extended to include such identifiers as "daffodil," so that anyone else who was a daffodil 7 was close kin, while, say, a daffodil 5 would be a more distant relative. As with real families, Vonnegut notes, you could tell your artificial relative to take his request for a handout elsewhere, but one imagines that would be less likely than if the request came from outside your artificial extended family. Artificial families among humans really do exist to some extent (unions, volunteer fire departments, religious groups), demonstrating, as in Vonnegut's model, just what a powerful role the concept of kinship (even artificially imposed) plays in our everyday thinking.

Another interesting means by which kinship and cooperation have worked themselves into our everyday decision making can be seen in the debate on inheritance taxes. Conservative politicians have argued that inheritance taxes should be reduced, if not abolished. Yet Irwin Seltzer, a member of a conservative think tank called the American Enterprise Institute, has noted something of a paradox in this logic.[32] How, asks Seltzer, can conservatives argue against affirmative action because it provides unearned benefits for a group of people and at the same time support inheritance tax cuts, which would also confer unearned benefits on a specific group of people? According to George Will, the answer is quite simple and there is no paradox in holding both these positions because "preferences administered by government and based on race are inherently obnoxious, whereas preferences based on kinship and administered by parents are not."[33] Simply put, kinship is just plain different from other categorizations we make.

Before addressing what we can learn about human cooperation from animal examples, let's take a moment to examine what we cannot learn from such scenarios. Consider the appalling act of infanticide. Male lions are known to kill the young in a group once they form a pair bond with a female.[34] This is not uncommon among animals; evolutionary biologists have long argued that males commit this act both to avoid devoting resources to children that are not their own and to bring females into reproductive condition more quickly. Revolting behavior by any standards, but in animals one thinks of it as an amoral act rather than an immoral one.

Yet, shockingly, the lion example can be matched by an equally scary figure for humans. The statistic arises from the work of Martin Daly and Margo Wilson, evolutionary psychologists who study human aggression in its most extreme form. Daly and Wilson are psychologists by training but use evolutionary principles as the foundation from which to make their predictions about the human psyche.[35] In their book *Homicide,* Daly and Wilson report that the rate of child abuse (which includes murdering children) in the United States is on the order of one hundred times higher for children living with one stepparent and one natural parent than for children living with two natural (biological) parents.[36] The numbers are equally disturbing in Canada. Further, poverty per se and a number of other possible confounding factors were ruled out as causes underlying their finding. No matter how Daly and Wilson sliced the pie, children living with a stepparent (which almost always meant a stepfather) were victims of child abuse much more often than their counterparts in homes with two natural parents!

What, if anything, should be the policy consequences of Daly and Wilson's finding? After all, their report, disturbing as it is, clearly shows the increased risks to children when stepparents are in the house and therefore might say something about the role of kinship, cooperation, and aggression. What do these data, driven by an evolutionary hypothesis about aggression and kinship, tell us? To begin with, we should not simply assume that stepparents are a danger to the lives of their stepchildren. To do so would be a grave injustice and precisely the way we should not be using evolutionarily inspired data sets. There are two reasons that any policy changes to emerge from Daly and Wilson's findings would be suspect (and Daly and Wilson themselves warn against deriving public policy implications from their work). First, despite the fact that the rate of child abuse is many times higher among stepparents than natural parents, the rate in both groups is very low in absolute terms; stepparents are more dangerous, but neither step- nor natural parents are *likely* to be dangerous in the first place. Second, if a single stepparent increases the risk to a stepchild, then, according to kin selection theory, two stepparents (neither of which is related to the child) should be an even more dangerous situation. Yet Daly and Wilson note, "We would not especially anticipate elevated risks to adopted children." When two individuals adopt a child, the rate of child abuse among such individuals is probably the same as in the case of two natural parents. This suggests that what truly matters in avoiding child abuse is two loving parents, regardless of their blood relationship to a child. Since so many of our legal systems assume that a child is always better off with a relative than with a stranger, perhaps we should find a way to guarantee that lawyers, judges, and legislators recognize the statistics regarding adoptive versus natural parents and perhaps rethink this issue.[37]

Given that we avoid the pitfalls inherent in the infanticide case, how might we best use our evolutionarily derived notion of the importance of kinship to promote and structure human cooperation? Rather than constantly enticing people to cooperate by providing some immediate benefit, we might structure advertising to show people how their relatives are likely to benefit from some action taken and how this can reverberate down generations. Surely, the belief that helping family is the right thing to do is to some extent ingrained in almost all human cultures, but we can use the animal cases we reviewed earlier to force us to pick specific scenarios and elaborate on them.

Politicians seem to have learned this lesson well. Washington politicians and pundits choose to argue not so much how the problems and threats facing America will harm *you,* but rather how your children and grandchildren will be burdened by such problems unless you start cooperating. For example, Ross Perot, presidential candidate in the United States in 1992 and 1996, while surely overstating the problem, claimed that our grandchildren will have three-fourths of their income siphoned off to pay taxes because the American government refuses to set policies in motion to lower the national debt.[38] So Perot and virtually all other politicians in power argue that cutting some government program near and dear to voters' hearts or raising voter taxes is in the best interest of their descendants. Such appeals have been made by politicians for a wide assortment of problems that they would like the public to solve in a cooperative fashion.

What other types of behavior might be more accessible if appeals were made based on the effects on kin? In principle, the answer is "all of them," but in practice some behaviors seem more prone to being affected by such arguments. One possible example is environmental awareness. Although some environmental groups use a "make the world a better place for your kids and grandkids" appeal, many rely on slogans urging people to make the world a better place, period. Certainly such appeals work to some degree, but a slight change of wording might have a big impact on membership. The Sierra Club's motto (listed on their World Wide Web page) is a perfect example: "Protect America's Environment: For Our Families, For Our Future." Appeals to the desire for a better world rely on conscience, but appeals to kinship do so to a lesser degree and may therefore work more effectively.

Appeals to act cooperatively to help your kin obviously are most effective when people are convinced that cooperative acts will truly do just that. One way to ensure this is to have kin living near you. Humans have lived in extended families for the vast majority of our time on this planet, and larger groups (towns and cities) are relatively recent occurrences. Policies that, for example, provide some incentive to individuals and groups that form neighborhoods consisting primarily of extended families might spark kin-based cooperation. Sacrificial acts for the neighborhood—crime watch, cleanup—then amount to acts that help your kin.

During a recent trip I took to present a research seminar at Cornell, I had the good fortune to talk for an hour with Stephen Emlen (mentioned earlier for his work on white-fronted bee-eaters), who had some specific suggestions on how to accomplish community cooperation. In fact, in discussions with public policy makers at Cornell and in Washington, Emlen raised the idea that in certain circumstances, evolutionary models of kinship (many of which Emlen himself has developed) suggest that financial incentives to keep kin, such as grandparents and grandchildren, in the same neighborhood may merit further investigation. It is worth noting that Emlen's work on cooperation and conflict in white-fronted bee-eaters was the spawning ground for his ideas about kinship and cooperation in humans. Emlen notes that "bird studies are valuable because they can provide us with a window through which we can more easily view the fundamental biological rules that govern social interactions within family groups. By looking within this window, we can gain insights into some of the noncultural factors that affect our own social behavior. . . . It is my expectation that this Darwinian approach to the study of animal family systems will prove useful in allowing us to better, and more objectively, understand ourselves as we prepare to enter the next century."[39]

Emlen's ideas on financial incentives may not be as far-fetched as they seem. Evolutionary psychologists have conducted many experiments on people's tendency to sacrifice for the good of others. They consistently find that in written questionnaires subjects say that they are much more willing to sacrifice for "others" when the others in question are relatives.[40] Admittedly, the results of experimental surveys tell us only so much about human behavior. Furthermore, there have been no systematic studies on rates of cooperation in modern communities that have a strong kin bond versus those that do not. Nonetheless, the strong kinship component to human cooperation suggests that we investigate how cooperation fares in modern communities as a function of how related their members are.

One possible drawback to structuring communities based on relatedness might be what I'll call the "Hatfield/McCoy syndrome." Here, cooperation within families leads to intense conflict *between* families. For example, when I mentioned the idea of kin-based neighborhoods to Monique Borgerhoff-Mulder, an evolutionary anthropologist at the University of California (Davis), she was quick to note that the pastoralist peoples of Africa (many of which Borgerhoff-Mulder has herself studied) live in communities that are structured around kin, and if anything, these communities often display more aggression and less cooperation than non–kin-based groups.[41] Obviously, cooperation is not necessarily the result of kin-structured communities. Perhaps Western societies are so dramatically different from the pastoralists of Africa that the lack of cooperation in kin-based societies in Africa is no indicator of what would happen in the Western world—we simply don't know.

Animal examples of kin-based cooperation might serve as a guide for our moral compass by suggesting not only appeals to the effect of cooperative actions *on* kin, but ways that we can facilitate active cooperation *between* kin. To take a rather violent case as an exemplar, consider thirteenth-century England. Despite the fears of modern urban Americans, the homicide rates in the most dangerous modern cities are less than those of, say, Oxford in the 1200s. Murders were all too common at this point in history, common enough that underlying trends can be detected. One statistic of particular interest compares the relationship of victim to offender with that among co-offenders. Kin were five times more likely to act together to commit a murder than they were to kill one another.[42] In other words, if dirty work had to be done and it required cooperation, kin were quite likely to work together as opposed to acting as adversaries. While we surely do not wish to encourage kin to cooperate in the act of murder, this example illustrates that kinship can facilitate cooperative actions, even under the most dire circumstances.

Given the fact that kin receive inclusive fitness benefits when cooperating *with* other kin, can both business and society structure certain activities to capitalize on this bit of biology? Let me raise a few possibilities to begin with, and then examine the slippery slope I have outlined. In institutions as diverse as the army and the police department—institutions wherein serious danger is an everyday threat and acts of cooperation may require bravery—might not some units be structured, at least in part, around relatedness? If we expect cooperation in the face of very difficult odds, why not recognize that kin, in most cases, are the most likely group to undertake such actions to benefit one another? Most soldiers are probably more likely to risk life and limb in an army unit containing a sibling. The question of whether altruistic acts of bravery were more common when kin were in the same U.S. military unit has, however, not been "systematically explored."[43] (Siblings are usually not placed in the same unit, unless the individuals themselves request it, for fear that a family would lose multiple members should hostilities break out.) As an evolutionary biologist, I find this lack of data fascinating, as this would be one of the first things I'd explore when examining questions of bravery.

There is, however, some indirect evidence from the evolutionary anthropology literature that individuals would be more willing to risk dangers if they were surrounded by their relatives. The Yanomamo people of Venezuela have earned the unenviable, but accurate, moniker of "the fierce people."[44] Violence plays a large role in Yanomamo social dynamics. Although much of this violence occurs between rival groups, within-group aggression is also quite common. With respect to the topic at hand—kinship and taking risks—the Yanomamo are a kin selectionist's dream come true. If you want to know who will line up behind each putative combatant in a group, what you need is a bunch of family trees. Yanomamo males will almost always back the combatant to whom they are most closely related. Of course, relatedness is not the only factor determining alliance formation, but it is the predominant one. While the Yanomamo example does nothing to prove that soldiers in Western armies would be more willing to risk life and limb if they had kin in their unit, it certainly suggests that this possibility is worth exploring more systematically.

Within the business community, kin selection thinking can be applied to the production of items that require a team of individuals to cooperate with one another in the production phase and for which commission is paid to members of such a group. The ideas remain to be tested, but I'd predict that when such groups are small and contain kin, individuals will work harder because of the kin-related benefits (money going to both you and your relatives). Companies might even be able to advertise toward creating such groups and pay individuals a higher commission, since they are likely to receive greater output from such groups. Robert Ford and Frank McLauglin, in an article for human resources professionals, note that there are three general arguments for allowing relatives to work together. These arguments, though not couched in the jargon of kin theory, clearly suggest the potential benefits of such a policy:

- Nepotism is good for the small family owned organization because it provides an efficient way to identify dedicated personnel to staff such organizations.
- Permitting nepotism allows considering all potential employees who might be effective contributors to an organization, rather than arbitrarily excluding a large pool of people simply because they are related by blood or marriage to an existing employee.
- Nepotism tends to foster a positive, family-type environ-

> ment that boosts the morale and job satisfaction for all employees, relatives and nonrelatives alike.[45]

In many ways, Japanese corporate "families" revolve around creating a scenario in which workers at a plant are treated as though they are "family." Perhaps it might be worth trying this with real relatives.

Let us return for a moment to some potential problems with this approach of creating kin-based units in either the military or business worlds. First, when intense cooperation is called for in response to dangerous situations, placing relatives in a group runs the risk of people acting in irrational ways to help kin, rather than nonkin, within the group. That is, we might take a greater risk to help kin than is truly merited, taking all things into account. Further, there is the ever-present chance that in business, kin might cover for one another and help themselves rather than the company they work for. This is always a risk for any set of employees, but even more so in this case since kin can gain more by such actions than unrelated individuals. Moreover, there is a risk that people will feel that less is expected of them when they are working for a relative. This is a slightly different matter from having kin working *together* in a group, but subtly related. Whether the benefits of policies that promote kin-based decisions outweigh the costs for institutions such as the army and police should be considered on a case-by-case basis, but kin-selected cooperation suggests that such an analysis is worthwhile.

I need to make a very important caveat here: as with every argument I make using animal examples and theory to guide our moral compass on how to be more cooperative, I am not suggesting that we implement any particular plan. Each idea I put forth has its pluses and minuses, and some of these have profound implications for society. Rather, I am arguing that we should focus on what we might learn about facilitating human cooperation from a large database (animal examples and sound evolutionary theory) in a way that we have not done in the past.

There is one additional area that kin selection thinking suggests may facilitate human cooperation. The concept is quite simple and it has been around in various forms, no doubt, as long as people have been capable of its construction: kinship can be used to diffuse aggression and increase cooperation among opposing groups. One excellent example is the marriage agreements that characterized many civilizations throughout history. Marriage across royal families from different countries not only increased the wealth of all parties but also often (though not always) decreased the likelihood of aggression between the two lands.

The logic undermining this approach can be applied in a wide variety of circumstances, not merely royal families and marriages. Potentially, any two groups can be brought closer together by creating kinship bonds between them. Marriage is certainly one path, but any legal means by which members of one group are transferred to another group, while their kin stay put, is likely to decrease potential aggression and increase between-group cooperation. Once again, this approach is not without its problems, as there are many cultures that have specific traditions banning marriage with those in other groups. The notion that anyone should encourage creating blood ties between such groups is offensive to members of these cultures.

Understood in the appropriate context, then, examples of kin-selected cooperation may prove useful to us in our attempts to increase the level and degree of human cooperation. It is important to stress, however, that kin selection theory does not suggest that we should withhold cooperation from unrelated individuals. Such a suggestion would be unfounded and ludicrous, based on what we understand of both human nature and evolutionary theory.

Kinship is not the only path to cooperation. In fact, as discussed in the Introduction, three of the four paths to cooperation do not involve family dynamics at all. Let us now take a look at the next route to cooperation.

2 One Good Turn Deserves Another

Do unto others as you would have them do unto you.

—The Golden Rule

The Golden Rule and its variants are often the first society-based ideas we try to instill in our children. Why? First and foremost, from the standpoint of a value system, "Do unto others as you would have them do unto you" is the Judeo-Christian imperative. The second likely reason people stress the Golden Rule is probably a bit more practical. Humans are remarkable scorekeepers; we know who did what to us and when. Acting in accordance with the Golden Rule, then, is often the most politically and economically astute thing to do, as it has the effect of raising one's reputation.

Although a bit unsettling, one theory on why reputation is so important (often worth dying for, in extreme cases) is that it is a means of convincing others that you abide by the Golden Rule, that you can be trusted in reciprocal transactions. This in turn makes you more likely to be chosen from a pool of many, when good opportunities are available.[1] For example, if someone is looking for a partner to let in on the ground floor of a great business deal, he is much more likely to choose an individual who has a reputation for cooperating with associates than one who lacks such credentials. Why should be obvious—how someone acted in the past is the only guide we have as to how she will act in the future! Reputation, of course, may not accurately represent someone's actual behavior. Gossip warps information. Just watch your children play the game "telephone" and see how different the end product is from the initial message. Then imagine that facts about someone's reputation are being transferred in the game. In the absence of any other clues, however, knowing about someone's reputation is often better than nothing.

Just as we are more likely to cooperate with those who have been nice to us in the past, so too are we often less inclined to cooperate with anyone in circumstances where we believe *future interactions* are uncertain. A personal example (of which I am not particularly proud) illustrates this. Back in 1989, while in the process of completing my Ph.D. at the State University of New York, I would often travel up to Cornell University to use their experimental pond systems.

The project I was engaged in at the Cornell ponds examined whether bluegill sunfish were capable of recognizing one another, and if so, how this ability structured their social interactions. While this project was under way, the manager of the experimental ponds asked me to do some simple cleanup work at my site each afternoon. I, of course, obliged, for without the pond manager's help and expertise, my experiment would never fly. When my experiment was complete, the manager asked me to do one last bit of cleanup that might have taken all of one afternoon. I distinctly recall thinking that my experiment was finished, I had done enough tidying up already, and what's more, I would probably never use the Cornell ponds again, nor see the pond manager in the future. I then proceeded to convince myself that I had better things to do with my time and simply skipped the last afternoon's worth of cleaning. In other words, I'd never need the pond manager to cooperate with me again, so I really didn't care that I was now on his blacklist. In my mind, scorekeeping was now completed, or at least irrelevant. By the way: (1) bluegills do recognize one another; and (2) I now feel sufficiently guilty and plan to track down the name of the Cornell pond manager and send him an apology.

Each of us can think of some similar examples in our own lives, both of preferring to interact with known cooperators and not cooperating when we thought we would never meet someone again. For example, are you more likely to tip low in a neighborhood restaurant or in one across the country? There is a surprising twist, however, to our scorekeeping abilities that is a bit more subtle than the situation discussed above. It seems that our scorekeeping abilities, though impressive, are in some sense restrictive, in that we are not good at keeping track of everything equally well. Rather, we are exceptionally good at keeping track of what Leda Cosmides and John Tooby refer to as "social contracts."

Cosmides studied this phenomenon using the Wason Selection Task Test, a diagnostic tool developed in psychology.[2] To see how this test works, try to solve the following problem: You have a new job and one of your responsibilities is to ensure that certain documents already collected have been processed correctly. Your job is to make sure that documents conform to the following rule: If a person has a *D* rating, then his document must be marked *3*. You fear that the secretary you replaced may not have categorized people correctly. The cards below have information about four people and each card represents one person. One side of the card gives a person's letter rating and the other side tells that person's number code. Indicate only those cards you definitely need to turn over to see if any of these documents violates the rule.

Which card or cards do you need to turn over? Answer: the *D* and the *7* cards. If the *D* card doesn't have a *3,* it violates your rule, and if the *7* card has a *D*, it also violates the rule. If you answered incorrectly, don't be too dismayed—only 25 percent of those taking the test answered this question right.

Now imagine that you are asked to take on the role of a bouncer at a bar whose owner tells you the following rule: If a person is drinking beer, then he must be over twenty-one years old. The cards below have information about four people sitting at a table in a bar. Each card represents one person; one side of a card has the person's age and the other side tells what beverage that person is drinking. Indicate only those cards you definitely need to turn over to see if any of these people violate the rule.

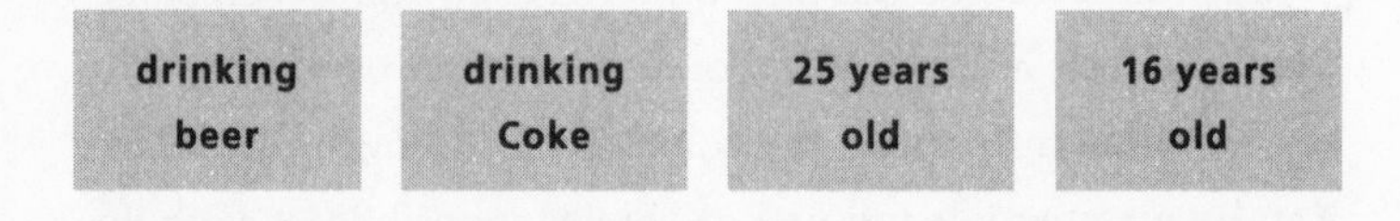

Which card or cards do you need to turn over? Answer: the "drinking beer" card and the "16 years old" card. Just as you shouldn't have been dismayed to get the first test wrong, don't get too ecstatic about getting the right answer this time—most people do. What makes these two tests so fascinating is that they require *exactly* the same type of logic to solve and yet we see such dramatic differences in the ability to find the solution. One can take both problems and think of them in the abstract as "If *P,* then *Q*" rules; in the first case, *P* is a *D* rating and

Q is the number *3,* and in the second, *P* is drinking beer and *Q* is being twenty-one years old. To solve an "If *P,* then *Q*" problem, one always needs to look at the *P* card (*D* in our first example and "drinking beer" in our second) and the not-*Q* card (7 in the first case and "16 years old" in the latter).

Why should we be so adept at keeping track of socially mediated events like the bar problem, as opposed to other things? Cosmides and Tooby have put forth the concept of a "Darwinian algorithm" to address this and similar questions.[3] Essentially, they argue that the human mind can be thought of as a conglomeration of computer-like programs (algorithms), each written for a specific function we must perform. Natural selection is then free to act on each algorithm independently of the others. In this schematic, we can envision one algorithm providing the rules for scorekeeping in social contexts (social contracts) and a different algorithm in place for scorekeeping in nonsocial circumstances. If selection favors social scorekeeping more strongly than other types of memory, then we can expect to find just what Cosmides uncovered in her study! In addition, the literature on the Wason Selection Task demonstrates that in virtually all cases where the problem is put in the context of social contracts, people do quite admirably; otherwise they fail miserably.

Our natural instincts for scorekeeping have spilled over into evolutionary approaches to the study of cooperative behavior, insofar as cooperation via reciprocity is one of the most active areas of research within behavioral ecology. Reciprocal cooperation has been examined in everything from worms to chimpanzees and in a wide variety of behavioral contexts. Before moving on to some of these incredible examples and pondering what they can and can't tell us about making humans more cooperative, let's take a look at the evolutionary theory underlying why reciprocity exists at all.

Why Not Cheat?

> *He who receives a good turn should never forget it: he who does one should never remember it.*
>
> Chinese fortune cookie

Evolutionary biologists attempting to construct a theory on how reciprocal cooperation could evolve are faced with a riddle. Why should an individual who has just been the recipient of a cooperative act ever reciprocate in kind? After all, natural selection is concerned strictly with costs and benefits and someone who cheats (does not reciprocate) receives the benefits of being a recipient without ever having to help anyone in return. Cooperators, on the other hand, receive benefits but also pay the costs of helping others in the future. On the cost/benefit worksheet, then, cheating appears to be favored; and so the riddle is not why we don't see more cooperation but why we see any cooperation at all.

There is truly only one solution to this riddle: somehow or another, under certain conditions, cheating does not pay better than cooperating. Trying to determine what those conditions might be, however, is not straightforward and has kept some of the best minds in behavioral and evolutionary ecology active for a long time. The first real attempt to understand the evolution of cooperation via reciprocity was undertaken by Robert Trivers in 1971.[4] Trivers's now-classic 1971 paper "The Evolution of Reciprocal Altruism" tackled the evolution of reciprocity by building on the already large experimental database available in psychology on the human tendency to reciprocate acts of kindness.

Recognizing that W. D. Hamilton had recently addressed altruism and cooperation among kin, Trivers set out to understand the evolution of reciprocity among unrelated individuals. As in my Chapter 1, "All in the Family," Trivers illustrates his ideas by the "drowning man dilemma." Imagine that a person is drowning and that if no one attempts to save him, there is a high probability that he will die. Further imagine that if someone attempts to save the drowning person, this Good Samaritan also incurs a risk of drowning, although a much lower risk. If this were an isolated event, and the putative Good Samaritan just calculated the costs and benefits of the act, he should not attempt to save the drowning individual. But suppose the two individuals in this scenario will meet each other many times again, and perhaps sometime down the road their roles will be reversed. Then a cost/benefit analysis might show that the Good Samaritan should try to save the drowning person, because this act is not all that costly and might be repaid with a huge benefit later on (when the person who was saved reciprocates).[5] In a nutshell, then, a cheater (one who decides not to help the drowning person) may pay a price for cheating and this price may be his or her own life. Under such conditions, cooperation via reciprocity can thrive.

Trivers's ideas on reciprocity, while stirring some controversy, were generally very well received. In fact, "The Evolution of Reciprocal Altruism," in conjunction with a handful of other papers, marked the start of modern behavioral ecology.[6] To many evolutionary biologists, however, Trivers's reciprocity paper lacked one critical element: a more detailed formal proof of his (predominantly) verbal arguments.[7] That proof came in 1981 when Robert Axelrod, a political scientist, and

William Hamilton, an evolutionary biologist (well known for his ideas on kin selection), forged a collaboration to tackle the question of the evolution of cooperation via reciprocity. Axelrod was already esteemed in political science circles for his creative approaches to studying reciprocity in a wide array of human scenarios and had recently published two path-breaking papers on reciprocity in a well-respected political science journal.[8] He believed, however, that an *evolutionary* approach to the mysteries surrounding human and nonhuman reciprocity would prove even more illuminating.

In an attempt to open a dialogue with behavioral and evolutionary biologists, Axelrod contacted Richard Dawkins (at Oxford University), one of the best-known and most respected evolutionary biologists in the world. Dawkins told Axelrod that the evolutionary biologist he really needed to talk with was Hamilton, who happened to be at the same institution as Axelrod—the University of Michigan.[9] Thus began the unlikely collaboration that is thought by some to have solved the riddle of the evolution of reciprocity via cooperation.

Axelrod and Hamilton recognized that a wonderful metaphor already existed for capturing the "cheating" problem inherent in cooperation. This metaphor, called the "Prisoner's Dilemma" game, was created and popularized by economists Jon Von Neumann and Oscar Morgenstein in the early 1950s.[10] The Prisoner's Dilemma begins with two crime suspects being interrogated by the police in separate rooms. Cooperation and cheating are defined from the perspective of the suspects. A suspect who cooperates doesn't "rat" on the fellow being grilled in the other room—for example, suspect 1 says suspect 2 did not commit the crime. Cheating is then defined as one suspect implicating the other.

For the sake of illustration, imagine that if one suspect cooperates and the other cheats, the cheater walks away free for turning state's evidence and the cooperator goes to jail for ten years, as all available evidence points to both of them. If both suspects cheat, they each go to jail for three years as they have implicated one another, but if both suspects cooperate (and don't squeal), the police have only minimal evidence, just enough to lock each suspect up for one year. The question then remains: should our suspects cooperate or cheat?

Let's begin to answer this question from the perspective of suspect 1. There are four possible outcomes for this suspect, summarized in the matrix (technically called a payoff matrix) shown here.

PRISONER'S DILEMMA FROM PERSPECTIVE OF SUSPECT 1

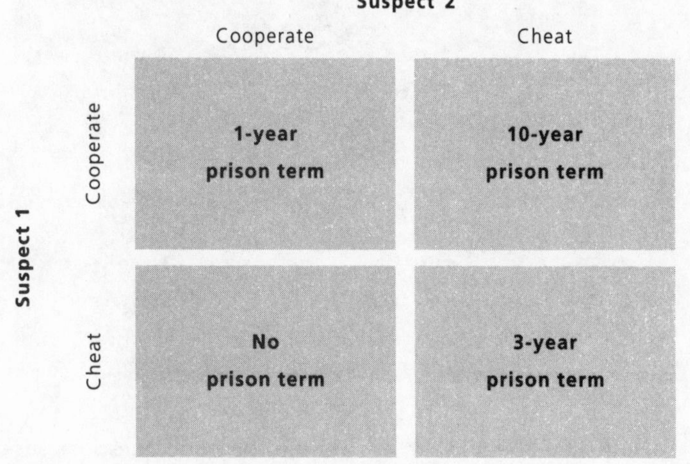

What should suspect 1 do, cooperate or cheat? One way for this suspect to decide would be to ask, "What would be the best thing for me to do if suspect 2 cooperated?" Answer: "Cheat, because if I cheat when suspect 2 cooperates, I walk away with no prison time, but if I cooperate I go to jail for a year." Now suspect 1 proceeds to ask, "What would be the best thing for me to do if suspect 2 cheated?"The answer again is "Cheat, because if I cheat when suspect 2 does this as well, I go to jail for three years, but if I cooperate when he cheats, I go to jail for ten years!" From suspect 1's perspective, then, regardless of what suspect 2 does, he should cheat (because it always produces a shorter sentence than cooperating).

What should suspect 2, in a separate cell and with no contact with the other suspected criminal, do? His payoff matrix, which follows, is identical to suspect 1's.

PRISONER'S DILEMMA FROM PERSPECTIVE OF SUSPECT 2

		Suspect 1	
		Cooperate	Cheat
Suspect 2	Cooperate	1-year prison term	10-year prison term
	Cheat	No prison term	3-year prison term

Suspect 2 should then ask himself the same questions suspect 1 did (What would be the best thing for me to do if suspect 1 . . . ?); he also decides that the only thing he can do is cheat. Now for the dilemma in the Prisoner's Dilemma. Both players opt to cheat and hence they each go to jail for three years. But, had both suspects cooperated with one another, they would have gone to jail for only one year each. *The best solution for each player leads to a suboptimal solution for both.*

Is there an escape from this dilemma? Is there some way to salvage cooperation? The answer is yes, and Axelrod and Hamilton did just that by approaching the Prisoner's Dilemma from a behavioral ecology angle. To begin with, Axelrod and Hamilton considered an iterated Prisoner's Dilemma game—that is, the game played many times between two individuals. This allows for behavioral rules that incorporate reciprocity, in one way or another. In addition, they conducted what they refer to as an "ecological" tournament, though in reality it was more of an evolutionary tournament.

The tournament was run on a computer and the rules were simple. Sixty-two individuals in fields as diverse as physics, biology, economics, mathematics, and political science were asked to send in a computer program that coded for any sort of behavioral strategy they wished. A program could mimic always cheating, always cooperating, cooperating every other move, cooperating every third move, or virtually any variant the programmer could produce. All these programs were pitted against each other (in a pairwise fashion) and the total score that each program received was calculated. At this point, strategies were then put into a "second-generation" tournament. How many copies of a given strategy were put into the second-generation tournament depended on how well the strategy did in the first tournament—strategies that fared well in the first round of the tournament were represented more than strategies that did poorly. This process was then repeated for many more "generations," simulating the process of natural selection.

This computer tournament, in conjunction with some mathematical analysis, showed that if the probability of interacting with a partner in the future is sufficiently high, one cooperative strategy fares very well, providing a possible means of salvaging cooperation from the Prisoner's Dilemma.[11] This strategy, submitted by Anatol Rapoport, a psychologist from the University of Toronto, was called "Tit-for-Tat" (TfT; this strategy has also been referred to as "The Judge"). TfT was one of the simplest strategies in the tournament; it instructs individuals to start off cooperating when they encounter a new partner and from then on to do whatever that partner does—if your partner cooperates, you do as well; if she cheats, you do likewise. In short, cooperate initially and then copy your partner's action. TfT is then a sort of "conditionally" cooperative strategy in that you condition your cooperation on that of your partner.

The reason TfT fares well in such tournaments is that it allows cooperators to continue to cooperate with others like themselves, thus reaping the relatively high payoff in the Prisoner's Dilemma when two players cooperate (see matrices), yet cooperators only get hustled once by cheaters (the first time they meet such individuals). Three properties of TfT seem to favor its evolution. Individuals employing TfT are (1) "nice" in that they always begin cooperating when encountering a new partner, which allows potentially cooperative individuals to start off on the right foot and avoid mistrust from the outset; (2) "retaliatory," as TfT players always respond to a cheater's action by cheating themselves; and (3) "forgiving," in that TfT players do not hold grudges but simply do what their partner did on their most recent interaction. So, for example, should someone cheat ten times in a row but then suddenly cooperate on play 11, an individual playing TfT will cooperate with such a partner after move 11. In essence, TfT individuals "forgive" prior cheating if a partner indicates he is willing to cooperate in the present. The forgiving nature of TfT appears to be important in that it allows individuals to make mistakes (e.g., cheat when they intended to cooperate) but subsequently slip back into a mutual cooperation mode.[12]

The nice, retaliatory, and forgiving characteristics of TfT make it a powerful force, so powerful in fact that rules guiding reciprocity can, in some cases, be more important than rules guiding cooperating with kin. For example, in a paper entitled "Why Be Nice to Your Rotten Brother?" my colleague David Sloan Wilson and I have shown that in a computer simulation designed to favor cooperating with kin, individuals who play TfT (and hence cooperate with cooperating kin but cheat on cheating kin) fare better than individuals who are unconditionally cooperative toward their kin.

The finding that TfT can outcompete other strategies that base their tendency to cooperate on kinship has implications for some of the debates currently surrounding what sort of values we wish people to associate with American society. While most people would agree that blood is indeed thicker than water, there is also a strong sentiment among many engaged in the debate over values that all that matters is whether people are good—not their race, religion, sex, or any other trait including whether they are or aren't your kin.[13] Nasty kin, the argument goes, are just as bad as nasty unrelated individuals, and one should judge others only by their actions. Interestingly, as we have seen above, evolutionary theory also makes a similar prediction in that such rules as TfT fare well even when they are pitted against rules that favor kin.

TfT's beauty is in its simplicity. It is really just a modern, technical version of the biblical "eye for an eye" concept: be nice, but respond in kind if people aren't.[14] Now that we see that a sound basis exists within evolutionary theory to explain cooperation via reciprocity, what evidence exists that living creatures are capable of this sort of behavior?

Playing Fast and Loose with Eggs and Sperm

"For Aristotle in *Historia Animalium,*" writes Patrick Colgan, "the motivation of fish ranged from enjoyment of tasting and eating to madness from pain in pregnancy," and for Francis Day, summarizing for the 1878 Proceedings of the London Zoological Society in the wake of Darwin's epochal *The Expression of the Emotions in Man and Animals,* fish were variously moved by "disgrace, terror, affection, anger and grief."[15] This quote is surely a grand overstatement of life as a fish. Still, even though fish are less socially complex than humans, they are probably much more complicated than most people give them credit for. In fact, many of the most carefully studied cases of cooperation via reciprocity are in this animal group. Consider the admittedly bizarre case of reciprocity in some hermaphroditic species of fish.

Since being male or female is one of the critical features that define our individuality, it usually comes as a bit of a shock to realize that many individual animals (and even more plants) are hermaphrodites—that is, a single animal can possess both male and female reproductive organs. There are two variants: sequential hermaphrodites, who are male at one stage of life and female at another, and simultaneous hermaphrodites (the group we will focus on), who possess male and female organs at the same time. Among the vertebrates, simultaneous hermaphroditism is relatively rare and occurs only in fishes, most often in deep-water species with low population densities.[16]

Before looking at reciprocity in hermaphrodites, we must bring out one of those nasty, unfair facts about sexuality: eggs are energetically expensive for females to produce, but sperm are cheap for males to make. Eggs are much larger and often filled with various nutrients, while sperm are essentially tiny bits of genetic material wrapped in a cover. Yet it takes *both* a sperm and an egg to produce an offspring and this sets up a rather fascinating problem for simultaneous hermaphrodites wishing to mate: why not always produce cheap sperm and have someone else invest in egg production? The offspring is just as related to you and yet you only need to produce a fraction of the resources that egg producers contribute! Of course, if most everyone takes this route (sperm rather than egg production) there will be virtually no matings! How do simultaneous hermaphrodites solve this dilemma? The answer is reciprocity.

In at least eight species of sea bass, we see the same solution to the "sperm are cheap, eggs are expensive" problem. Potential mates pair up and shed sperm and eggs into the water. But rather than having one individual produce all of the sperm for the pair and the other contribute the eggs, both individuals parcel out their eggs. That is, they divide up all the eggs they have into smaller packages, and on any given turn one individual will shed sperm and the other will shed one packet of eggs. The individuals take turns—if you gave eggs last time, you shed sperm this time and vice versa. Eric Fischer has studied this phenomenon for many years and describes the mating process of simultaneous hermaphrodite sea basses: "Individuals form pairs late in the afternoon. Before spawning, two mates alternate courtship displays, and the last fish to display releases eggs, while its partner releases sperm (fertilization is external and eggs are planktonic). Each fish releases only part of its clutch during a given spawning act, and partners regularly alternate release of eggs."[17]

Fischer and others have found that individuals alternate

roles (sperm producer, egg producer) 80 percent of the time. In essence, parceling eggs has forced sea basses to engage in cooperation via reciprocity. If eggs were not parceled, all individuals would choose to be male, but since you can only fertilize a small number of eggs on any given mating in the sea bass system, you need to cooperate and produce eggs if you want a crack at the next bunch of eggs your partner is going to produce. Otherwise, if you consistently produce sperm, your partner can simply find another mate who plays by the rules. In fact, the data suggest that individuals in one of the better-studied species, *Hypoplectrus nigricans,* fertilize, on average, about as many eggs as they parcel out.[18]

Some evidence exists that simultaneous hermaphrodites that engage in egg swapping actually use the tit-for-tat strategy outlined earlier. For example, black hamlets *(Hypoplectrus nigricans)* and chalk bass *(Serranus tortugarum)* may retaliate against cheating, just as the TfT strategy suggests they should. In these species, mates waited significantly longer to parcel out eggs to a partner who had failed to reciprocate in prior interactions.[19]

If it seems hard to believe that such a complicated system of reciprocity could evolve in a lowly fish, then data that a remarkably similar form of egg trading and reciprocity occurs in the hermaphroditic worm *Ophryocha diadema* should strike you as utterly fantastic.[20] What makes the worm example so fascinating is that as simple a creature as a polychaete worm is able to assess whether its partner is cheating and then respond appropriately!

A Nod in Your Direction

Interesting as those hermaphroditic cooperating fish (and worms) are, watching reciprocity in the context of egg parceling might not be all that exciting. For the stuff that TV documentaries are made of, we need to move to the Gombe National Park in Tanzania.

Imagine you have a protected seat high in one of the rare trees that speckle the savanna of Gombe. You happen to be watching a troop of baboons when you see the oddest thing. It looks like one male baboon is threatening a second male, while at the same time trying to get yet another male to come over to back him up. Surely this is an isolated incident, perhaps a result of the same luck that landed you up in the trees, safe from all danger in the first place. Not so, according to Craig Packer.

In one of the best-known studies of coalitions among primates, Packer studied male reproductive coalitions in baboons, *Papio anubis.* One of the aims of this study was to separate kinship from reciprocity, and while Packer could not be sure of the exact pedigrees of all individuals, known bloodlines and demographic data (males always emigrate from their home troop) suggest that coalition partners were unrelated.

Pairs of baboon males, Packer observed, actively solicit coalition partners in an attempt to gang up against a third male and gain mating access to the female he is guarding. This sentence may set off some alarm bells—it is one thing to suggest that animals reciprocate acts of cooperation but another to suggest that they "solicit" anything, let alone a coalition partner. So let's be as specific as we can and use Packer's own definition of soliciting: "a triadic interaction in which one individual, the enlisting animal, repeatedly and rapidly turns his head from a second individual, the solicited animal, towards a third individual (opponent) while continuously threatening the third. The function of headturning by the enlisting animal is to incite the solicited animal into joining him in threatening the opponent."

Access to a reproductively active female is a rare commodity in the baboon world, and males fight vigorously for it. Consequently, the only way that relatively smaller males can obtain access to a female is to join forces with another male, close ranks, and try to force a dominant male to let them near his mate. Packer observed ninety-seven coalitions form, and six of these resulted in the female leaving the dominant male and consorting with one coalition member. But not just any coalition member—always the individual that had enlisted help. Clearly, then, there is an advantage in enlisting help from others (possible mating opportunities), as well as a potential cost (chance of injury when interacting with the dominant male). Why, however, should any baboon respond to a solicitation to enter into a coalition? After all, the solicited individual faces the same costs as the solicitor but never gets the benefits of consorting with a female. It seems that selection would favor asking for help but never providing it when others do.

The answer to this riddle, not surprisingly, lies in reciprocity. Males who join coalitions when they are solicited are much more likely to have potential coalition partners respond to their solicitations in the future (it also turns out that dominant males are not likely to fight against a coalition and so the cost of injury to either member of a coalition is low). To show reciprocity, though, one must demonstrate more than this; a true buddy system must be in place. That is, it is not enough to show that those that joined when asked were likely to have others join when they themselves put out the request. One must rather show that "if I help you, you help me." Sure enough, Packer found just such a relationship in that individuals had "favorite" partners, and favorite partners solicited each other more often than they solicited other group members. Natural selection then seems to favor coalition formation in the long run:

> During the reproductive life of a male, which extends over 10 yr, there may be a large number of situations where it would be advantageous to enlist a coalition partner in an encounter against a consorting male. The number of offspring that a male sired as a result of participating in reciprocating coalitions would be greater than if he did not, while his life span would probably not be appreciably shortened by aiding coalition partners.[21]

In some ways it should not surprise us that primates, our closest biological relatives, engage in such cooperative coalitions when the benefits seem to swamp the costs. We humans certainly do—a vivid example at the international level being the Gulf War coalition formed in the early 1990s when Iraq invaded Kuwait. The Persian Gulf seems to be a popular place for coalitions; only a few chapters into Genesis, Abraham describes a war between two coalitions in this area.[22]

Yet despite the fact that coalitions at all levels are integral to human behavior, there is something vaguely eerie about the fact that animals also behave this way. The notion that baboons are complex enough to use other baboons as tools to further their own ends is just not what most of us think of when we picture primates.[23]

The Benefits of Ticking Off Your Fellow Impala

Let's stay in Africa and spend some time watching herds of impala prance about on the Masai Mara Reserve in Kenya. Impala, like so many creatures, face a problem that seems odd to us two-armed humans—they cannot reach many parts of their bodies to clean themselves off! Imagine trying to clean off your back if you were walking around on four respectable legs and you'll begin to see the problems so many animals face. Yet parasites and other nasty things that cling to skin virtually dictate that these hard-to-reach places be somehow cleaned. But how? One possibility is to have another impala do it for you. They can surely reach your back and neck and get rid of all the debris. But that never-ending problem resurfaces yet again: if someone cleans me off, why not just scuttle off and avoid the time and energy it takes to return the favor? Impala seem to have solved this problem much the way our hermaphroditic fish solved their egg/sperm dilemma—by parceling out the wanted resource (in the case of impala, bouts of cleaning). This in turn forces reciprocity on the part of the individuals involved in an interaction.

Benjamin and Lynn Hart have been studying grooming behavior in impala since the 1980s. They found that bouts of reciprocal grooming typically occur after one impala begins grooming a nearby neighbor. Grooming usually involves running one's tongue along the neck of a partner, about half a dozen to a dozen times. The recipient of such grooming then responds in kind. But one such bout of grooming is not sufficient to clean off an individual; the entire operation takes four to twelve bouts. Reciprocity emerges, as an impala simply cannot get groomed sufficiently unless it responds in kind along the way.

Grooming itself appears to carry some costs (energy expenditure, loss of saliva, decreases in antipredator vigilance, etc.), but the benefits seem to swamp such costs, as grooming may reduce the tick load of the individual groomed, with significant health consequences (just a few ticks can have very serious effects on weight and competitive abilities).[24]

Impala are remarkable scorekeepers when it comes to keeping track of grooming events. Whether pairs are male/male, female/female, female/male, subadult male/subadult male, or fawn/fawn, individuals almost always receive about the same number of grooming bouts that they hand out. The fact that reciprocal grooming is seen in fawns as young as three days old, and that young deprived of grooming exchanges with adults still display this behavior, suggests, but is not evidence for, a genetic component to reciprocal grooming in impala.[25]

If you think that cooperation during tick removal is a bit revolting, nature has much worse to offer.

Vampire Blood Money

It is time, once and for all, to set the record straight on the poor misunderstood vampire bat. Bram Stoker, Count Dracula, and images of evil aside, vampire bats are in fact quite social creatures and very cooperative, at least with respect to the thing they care about most—blood. People need to realize that life as a vampire bat is not easy. Vampires starve to death if they fail to get a new meal of blood every forty-eight to sixty hours, and this is not an unusual occurrence. In fact, this need for constant blood meals has produced one of the most fascinating examples of reciprocity in all of the behavioral ecology literature.

A typical group of vampires, living for example in hollow trees in Costa Rica, contains about eight to twelve females. Gerald Wilkinson has been studying such groups for at least fifteen years to understand why long-term association patterns between vampire females emerge, with some individuals living together for over three years. Wilkinson considered many hypotheses, but the data showed that one factor was the overwhelming reason for long-term associations: females regurgitate blood meals to each other when a roosting bat has failed to get its own meal and risks starvation.[26]

Three pieces of evidence suggested to Wilkinson not only that vampires regurgitated blood to one another but that reciprocity played a critical role in determining who got a blood meal when she was hungry and who didn't. First, in accordance with tit-for-tat models, vampire bats had a reasonably high probability of interacting with roostmates for long periods of time, thus allowing a time frame within which reciprocity could operate. Second, Wilkinson had evidence that vampires were capable of recognizing one another and remembering past actions, providing the cognitive machinery one needs for reciprocity in the vampire system.[27] Last but not least, in accordance with Trivers's idea on reciprocal altruism, obtaining a meal (the benefit) is extremely important, while providing some blood to a starving roostmate probably has only a minimal cost.

Sure enough, when Wilkinson examined blood sharing in detail, he found that individual vampire bats were much more likely to give a blood meal to bats that had provided such a resource to them when they were starving.[28] But Wilkinson did more than observe these associations; he ran a fascinating experiment. He collected bats from two different locations, put them together, starved various animals, and looked at the exchange of blood meals to test for reciprocity. Four adult female bats were taken from one tree roost (site 1) and five vampires—three adult females, an adult male, and an infant—were caught at another site fifty kilometers away (site 2). The four females from site 1 had lived together before this test, as had the three females and the infant from site 2. The male from site 2, however, presumably had no prior interaction with any of the females from his site (i.e., he knew no other vampire bats in this experiment).

Once the vampires from the two sites were housed together, Wilkinson systematically withheld blood from one individual each night, until each individual had been starved twice. What he found was that only bats that were on the verge of starving (i.e., would die within twenty-four hours without a meal) were given blood by any other bat in the experiment. But, more to the point, individuals were given a blood meal only from *bats they already knew from their site.* Site 1 females only fed Site 1 females, Site 2 females only fed Site 2 females and the Site 2 infant, and nobody helped out the poor male. Furthermore, vampires were much more likely to regurgitate blood to the specific individual(s) from their site that had come to their aid when they needed a bit of blood.[29] Granted, there is only so much one can glean from an experiment with one data point (as we have here), but the fit of these data to models of reciprocity is significant.

Floating Around Aimlessly and Causing Trouble

The vampire bat example shows the high stakes involved in games of cooperation via reciprocity. If you are labeled as a cheater and find yourself starving, you are on your own. Sometimes these high-stakes games can be dangerous for family members as well. To see this, let's look at the phenomenon of "floater males." In many breeding species of birds, there are only so many nests available. Some males in reproductive condition find themselves without suitable nesting grounds, thus all but guaranteeing that they will find no mates that year. In such a situation, males "float" from territory to territory, surveying the landscape for future mating sites and occasionally causing havoc for those lucky enough to have a successful nest.

At first glance it seems patently obvious that territory owners should always keep floaters away. At best, floaters are scouting out your nest to use in the future; at worst, they are an immediate threat. But in fact floaters are tolerated by territory owners because they are known to help out when nests they are examining are threatened—quite useful should a nest full of chicks attract a predator. Yet if floaters act aggressively when predators are not around, a nest owner's vulnerable offspring are the most likely victims. Michael Lombardo studied floaters in tree swallows *(Tachycineta bicolor)* and has argued that both breeders and floaters appear to be trapped in a Prisoner's Dilemma and might opt to play TfT.[30]

Lombardo first simulated an act of cheating on the part of nonbreeders: he surreptitiously placed stuffed models of floaters on a breeder's territory and then recorded the behavior of territory holders. Initially, territorial couples did not typically respond aggressively to the models. Then Lombardo removed two live chicks from the nest and replaced them with two dead chicks (Lombardo himself didn't kill any chicks). In removing chicks, Lombardo was trying to mimic intense cheating on the part of floaters.[31] When parents discovered this horrendous turn of fate, they responded swiftly and strongly, chasing nonbreeders off their territory; thus cheating on the part of floaters was followed by expulsion from a territory. In what seems like a rather fine-scale, subtle move on the part of territory holders, floaters were not chased if prior knowledge precluded them as the killers.

One unsolved mystery in Lombardo's tale of tree swallows is what possible payoff could ever make up for a floater killing two chicks. After sufficient time, territory holders apparently "forgive" floaters for their act of killing. Could it be that having a floater around to help protect the nest against future attacks is so important that it merits letting a known killer back in the area?

Guideposts

While writing this chapter, I had a chance to visit the zoology department at the University of Toronto. At a small dinner gathering I made a light comment to the effect that my wife and I had recently moved to a new area, that we had found an honest and skilled mechanic, and that such a find was so rare that we would in all likelihood never move again. In response, one of the dinner guests noted that when he looked for a mechanic, he always stayed away from garages on big highways and near "strips." Such mechanics, he said, knew that they were never going to see you again and were notorious shysters. Go to a neighborhood garage, where word of mouth serves as advertising, and they know that you will be a long-term customer. Reciprocity shapes all aspects of our decision-making process (and need not be as mundane as choice of mechanics).

As vampire bats, impala, hermaphroditic fish, and coalition-forming baboons demonstrate, cooperation via reciprocity sometimes manifests itself in bizarre situations. This is also the case for humans. One might, for instance, guess that enemies engaged in battle would be among the least likely candidates for some sort of system of reciprocity. But Robert Axelrod, in *The Evolution of Cooperation,* illustrates that this is far from true. Axelrod outlines the "live and let live" system of reciprocity that developed during trench warfare between the Germans and the British during World War I.[32] The war was notoriously bloody, in part because armies did not fully understand how the advanced weapons available would change the notion of modern warfare. As a result, opposing armies' units would often "dig in" and the same units would fight each other in trench warfare for months on end. In this horrendous setting a fascinating form of reciprocity emerged. In direct violation of orders from their respective commanders, troops on both the British and German sides would often choose not to aim their weapons directly at their foes. Rather, they would shell an area close to the enemy, to show that they could do harm whenever they so desired.

The troops in the trenches realized that they were going to be pitted against one another for long periods of time, and an implicit agreement was reached that they would not try to kill each other every day, all day; it was simply too dangerous. As long as both sides abided by this unspoken agreement, neither would suffer terrible casualties constantly. In fact, this agreement was not always unspoken—consider the following astounding recollection of a British officer:

> I was having tea with A Company when we heard a lot of shouting and went out to investigate. We found our men and the Germans standing on their respective parapets. Suddenly a salvo arrived but did no damage. Naturally both sides got down and our men were swearing at the Germans, when all at once a brave German got on his parapet and shouted out "We are very sorry about that; we hope no one was hurt. It was not our fault, it was the damned Prussian artillery."[33]

In such circumstances, the offending army would then even "accept" the enemy responding in kind without further retaliation.

Whether this sort of cooperation is something we wish to foster is certainly open for debate. Some will think of it as a crime deserving court-martial or even summary execution, while others may be more merciful and militarily subversive. Nevertheless, whatever your take on the "live and let live" system, cooperation via reciprocity can have a significant impact on human social dynamics.

Social psychologists have examined human reciprocity in virtually every possible scenario one can imagine.[34] There are thousands of these experiments, but the vast majority do not even mention evolutionary perspectives on reciprocity, let alone what animal studies in this area can tell us about how to make ourselves more cooperative. Before raising some ideas on that front, let me start out with a caveat. Even when cooperation via reciprocity is favored, we must be wary of who is cooperating with whom and for what means. Consider the baboon coalitions we reviewed earlier. Clearly males are engaged in reciprocal coalitions, and natural selection is favoring some baboons being nice to other baboons. Now switch to the perspective of the dominant male, whose consortship with a female is being threatened. In his eyes this whole coalition business is hardly cooperative, but rather aggressive. Cooperation and niceness are in the eyes of the beholder in this example. Fostering cooperation via reciprocal coalitions often selects for such coalitions to join forces against others; as such, it is a powerful tool and like all powerful tools must be carefully monitored. Remember, Axis alliance members in World War II were very cooperative with each other.

What can we learn from evolutionary theory and animal examples of cooperation via reciprocity? Keep in mind again that when we answer this question we are not, by any stretch of the imagination, trying either to copy what animals are doing or to do the opposite. Rather, we are using animal behavior as a sort of baseline model for what our actions might look like if we did not possess the elaborate cognitive machinery that defines humanity. Once we understand animal cooperation, we can then focus our moral compass on ways to foster human cooperation.

Two themes underlie many of the examples of cooperation that we have uncovered among animals, as well as much of the theory developed to understand animal cooperation. First, cooperation is fostered by the expectation of many future interactions with a partner, particularly when one is never quite sure when the last interaction with a partner will be. Second, parceling out benefits over time favors cooperation as it induces parties to cooperate continually to get the full value associated with an interaction. Let's look at each one of these in turn.

How can we better encourage cooperation, armed with evidence that evolutionary theory and animal examples point to such an important role for future interaction among potential cooperators? One obvious place to begin might be with occupations in which cooperation via reciprocity seems critical and long-term interactions within pairs of individuals is at least plausible. Occupations such as police officer or firefighter jump to mind. Common sense might seem to dictate, for example, that cooperation via reciprocity in police officers might best be fostered by keeping pairs of officers together for long periods of time so that they can learn about each other and achieve some level of mutual comfort. Evolutionary theory and animal examples do provide a subtle distinction here that might be important. What fosters reciprocity is not so much numerous past experiences with a partner and achieving a level of mutual comfort as it is the probability of future interaction. Remember, theory and some data suggest that strategies like tit-for-tat work not primarily because of a long series of past interactions (in fact, only the last interaction matters for this strategy), but because of the possibility of many chances to reap the benefits of cooperation in the future.

Can the subtle distinction between long-term bonds that focus on the past versus the possibility of future interaction really amount to anything but an academic exercise? Consider the following scenario: Bob, age 55, and Tom, age 35, have been patrol partners for the last ten years. Fred, age 25, is a new police recruit with glowing recommendations. Bob plans to retire in a year and the police chief has to decide whether to keep Tom partnered with Bob or to pair Tom up with Fred. What should he do? If he believes that future expectations are more important than past interaction, Fred and Tom should be partners, but if he thinks that a long history of interactions is critical, then Bob and Tom should remain a team.

I am *not* suggesting that police officers are motivated to cooperate strictly by whether they believe that sometime in the future their partner will "pay them back" with an equally nice action. Integrity and valor alone explain much of the cooperation we see in the police. But, given all the evidence that future interactions are important in fostering reciprocity, could it hurt to see whether some useful hints toward police policy might not emerge from such evidence? As with all the suggestions I make based on theory and animal examples, my overarching aim is to make readers think about cooperation from a different perspective. If such an exercise leads to increased cooperation, wonderful, but as long as it doesn't make things worse, we are the better, I would argue, for having explored all options.

Anyone with kids knows that they can sometimes be nasty to each other (not to mention parents). How could we strengthen the belief in our kids that the other children they are playing with now will be their long-term play partners, thus fostering reciprocity? One possibility is to lower class size in schools (and there are other, more obvious benefits to this as well) *and to keep classes intact from year to year*—children together in grade 1 can expect to be together in grade 2, grade 3, and so on. Such a system favors cooperation because of the possibility of reciprocity in the future. Of course, there are drawbacks to this means of potentially favoring reciprocity. Arguments could be made that children gain much by having new classmates each year and experiencing many different types of friends. Furthermore, while such acts might truly foster cooperation among children within a class, they also increase the chances that children in different classes will view each other as outsiders.

Of course, just as bravery might underlie much of police cooperation, so too might many other factors beside the possibility of future interactions underlie reciprocity in our children. Yet reciprocity plays an important and largely ignored role here—after all, reciprocity is part of the reason we feel so strongly about teaching our children the Golden Rule.

The second clear message emerging from the earlier part of this chapter (particularly the fish and impala examples) is that parceling benefits into numerous discrete packages favors reciprocity by creating the impetus for both parties in an interaction to stick around. Because it does not allow one party to "hit and run"—obtain the benefits but fail to pay any costs—parceling can be a powerful stabilizing tool.

Parceling benefits is fascinating from both the evolutionary and the psychological perspective and may provide some new insights in economic decision making. Many products can be purchased on some type of payment plan, where the benefits that a business receives (the money flowing in for a product) are delivered in parts (the payments), rather than in one large sum. Why such a parceling system? One obvious reason is that businesses often obtain an extra bonus—interest payments—in such a scenario.

Parceling also favors reciprocity between business and client and generates continuing pressure to produce a good product. This is illustrated most clearly in service industries. A housekeeper is more likely to be consistently conscientious if she is not paid in total up front. As with any service, there are many corners that could be cut; parceling payments makes this skimming much less likely, because such cheating can be punished.

The most fascinating aspect of this new perspective on accepting payments in part is that in some businesses, one could request full payment at the start and remove the pressure to perform at the highest level. Yet businesses often choose not to do this. Why? One reason is that reciprocity and high performance, over the long run, are more profitable than cutting corners. Thus in parceling, businesses have constructed a type of unconscious self-monitoring system. Such monitoring systems, however, are typically favored when a unit (in this case a business) or some section of a unit (say a housekeeper working for a service company) might cheat if the self-monitoring system were not present. One might argue, therefore, that businesses, at some level (even a level not recognized by the companies themselves), are aware of the ever-present temptation to cheat and hence *force themselves* not to succumb by creating a parceling system.

Evolutionary and behavioral ecologists find such a self-monitoring system remarkable. Back in the late 1970s, Richard Dawkins and John Krebs suggested that animals and even humans, under some conditions, may be programmed by natural selection to deceive, when deception provides a net benefit.[35] Dawkins and Krebs, among others, have further suggested that natural selection may operate to make individuals unaware that they are deceiving others. If not recognizing that you are deceptive makes you more likely to be deceptive and if this provides benefits, then selection can favor psychological mechanisms that shield individuals from even recognizing their own behavior as deceptive. The parceling example above, however, adds even another layer of complexity to this equation, as it suggests that the converse is also true. Natural selection can sometimes favor self-monitoring systems that make you choose a path that forces you to be cooperative, when you probably would cheat without that monitoring system—and you may not even recognize that you are monitoring yourself for that reason.

Overall, reciprocity is the best-studied evolutionary path toward cooperation, in both humans and animals. In that sense, this type of cooperation holds great potential, in that the better we understand something, the more likely we are to decipher what favors it and what doesn't. Yet cooperation via reciprocity centers on the notion that you get "payback" for nice deeds and that you need to do a lot of scorekeeping to make sure of who owes you what. There is, however, a pathway to cooperation that does not require such detailed cognitive abilities and simply entails cooperating when it pays and cheating when it doesn't. This "selfish cooperation" has been dubbed "by-product mutualism."

3 What's in It for Me?

Either God is or he is not. But to which view should we be inclined? Reason cannot decide this question. Infinite chaos separates us. At the far end of this infinite distance a coin is being spun which will come down heads or tails. How will you wager?

Let us weigh up the gain and loss involved in calling heads that God exists. Let us assess the two cases: . . . There is an infinity of infinitely happy life to be won, one chance of winning against a finite number of chances to lose, and what you are staking is finite. That leaves no choice; wherever there is infinity, and where there are not infinite chances of losing against that of winning, there is no room for hesitation, you must give everything.

—Blaise Pascal, *Pensées* 418

Some people, for lack of a better term, might be called "Pollyanna cooperators." These individuals believe that we are inherently good and we do good for good's sake, and, for certain types of cooperative behavior, they may not be all that far from the truth. Linnda Caporael and her colleagues, for example, have found that people often (but not always) cooperate when all egoistic motives for such cooperation are removed.[1] On the flip side of the coin, there are others—the "selfish cooperators"—who possess rather bleak views on why we (as well as nonhumans) cooperate. Consider the dilemma of two people who are trapped in a cave with a huge boulder blocking the way out. To move the boulder, they must act in coordination and so they do. They cooperate because they must; otherwise both are in worse straits—immediately. This kind of cooperation or teamwork can be generalized to suggest that we simply calculate the at-hand costs and benefits associated with a potentially cooperative act. If the benefits outweigh the costs, we cooperate; otherwise the hell with it. Taken to extremes, this view produces some bizarre and fascinating arguments about human behavior.

Blaise Pascal was a brilliant French mathematician and philosopher of the seventeenth century. In one of his more famous works, *The Wager,* Pascal outlines a cost/benefit analysis of whether one should accept God as the supreme deity. Pascal's basic argument is captured in the epigraph to this chapter. It is strange that Pascal wrote that reason cannot determine whether God exists, because reason seems to be exactly what he uses in his argument for belief. In essence, Pascal makes the very rational claim that nothing can outweigh the *infinite* cost associated with making the wrong choice in this process and so the cost/benefit worksheet always comes out in favor of accepting God.[2] Add in the potential benefits of choosing correctly—infinite benefits in the world to come—and the decision is obvious. In fact, skeptics sometimes assert that the only reason for people like the late Mother Teresa is that such individuals weigh the benefits of the afterworld quite heavily and are just making a cost/benefit decision to be hyperaltruistic in this world for payoffs in the next. Mother Teresa, the argument goes, was making an economic decision; she was just weighing costs and benefits differently than others. I don't subscribe to this view, but it is a reasonable hypothesis, like it or not. The trouble comes in testing this: How can we know what Mother Teresa's motives were? And even if we did know, why would it matter? These are difficult questions and ones that we may never resolve.

Pascal's cost/benefit analysis was not just an academic exercise. He accepted his own reasoning and gave up life as a philosopher and mathematician in favor of being a missionary. A fascinating recent parallel can be found in the late George Price, a brilliant engineer turned evolutionary biologist. Price made numerous fundamental contributions to the study of the evolution of behavior. Despite this, his life "ended in great sadness and poverty . . . tormented by the feeling that he failed to contribute in any significant way to easing human suffering."[3] Price devoted his last years to the study of religion, gave away all his earthly possessions, and went to live (and die) on London's skid row.

The idea that we cooperate when it is in our best interest, and only when it is so, has been advocated by Richard Dawkins in his books *The Selfish Gene, The Extended Phenotype,* and *The Blind Watchmaker*[4] and has been recently extended directly to the question of human altruism and cooperation in Matt Ridley's *The Origins of Virtue.*[5] The "selfish gene" view of the world holds that genes foster cooperation when it pays to cooperate. This then explains everything from the existence of individuals (which are coordinated clusters of genes acting in their own best interest) to the evolution of trade (which itself can be boiled down to genes looking out for themselves, just in rather complex ways). So, the notion that individuals cooperate when it is beneficial in the here and now and forgo this option otherwise is not new to evolutionary biologists.

Although the idea has been masked under different names in the past, the current moniker for this path to cooperation is "by-product mutualism." The name derives from the argument that mutualism and cooperation in certain cases are just "by-products" of the selfish interests of the participants. Two people each choose an action based on their cost/benefit analysis and sometimes, as a by-product of this decision, you get cooperation. Our cave dwellers trapped by a boulder could care less about anything but themselves, but they still cooperate. Yet, as with all paths to cooperation, there is some real theory underlying by-product mutualism and we need to become familiar with it, to some extent, before we can move on to animal examples and human applications.

Boomerangs and Harsh Environments

Scientists understandably prefer to have some solid mathematics supporting the theories they advocate. We have seen that experimental work on reciprocity and kin-based cooperation was in many ways driven by theory. Some ideas, though, seem so intuitive, so fundamental, that we really don't need all that much in the way of mathematical models to make broad-scale predictions. By-product mutualism may be one such idea. When coining the phrase back in 1983, Jerram Brown, though well known for his theoretical (and empirical) work in the area of cooperation,[6] captured the essence of this form of cooperation in two sentences:

> In by-product mutualism, each animal must perform a necessary minimum itself that may benefit another individual as a by-product. These are typically behaviors that a solitary individual must do regardless of the presence of others, such as hunting for food.[7]

Consider Brown's case of hunting (or for that matter, any manner of obtaining food, other than the grocery): we need to get food to survive and so we hunt when no one is around; otherwise we starve. Sometimes others are around and in some cases hunting with them gets us more food, so we hunt in groups. Cooperation is a by-product of our single-minded decision to get as much food as possible, as soon as possible.

We can, however, be just a bit more specific about how by-product mutualism operates. Let's imagine two general types of environments, "harsh" and "mild."[8] Harsh environments are tough to live in—there might be lots of predators, little food, food that is hard to obtain alone, or any other number of factors that might make life more difficult. Mild environments are the converse. It is important to recognize that the two "environments" here need not be thought of as separate physical places. Time is also a factor, and so a single physical space can be "harsh" in one time period and "mild" in the next. One can actually use this type of depiction of the environment to create a mathematical parameter that measures "harshness" and assign various types of environments a number. Of course, actually assigning a specific value of harshness to a particular environment will always be somewhat subjective, but in principle it is possible.

In our model of two different environments (harsh and mild), individuals must make many decisions about whether to cooperate with others or not. Their decisions depend critically on what has been labeled the "boomerang effect"—the chances that not cooperating will have an immediate negative consequence (i.e., effect on fitness) that outweighs any associated benefits of cheating. Harsh environments, the theory goes, are more likely to produce such boomerang effects and hence are more conducive to cooperation via by-product mutualism. In mild environments, it often pays to cheat or to just go it alone and skip group cooperative acts. In other words, cooperate when it pays, otherwise spend your time on more useful things. Experimental work on animals suggests that by-product mutualism is all too common in the natural world (as well as in the laboratory).

That's a Big Zebra—I Could Use Some Help

Examine the growth of ecotourism and related industries and you will come to an odd realization: just getting a chance to view nature has become a marketable commodity. People pay real money to observe the nonhuman world, and some ecologically minded economists have suggested that this may be the saving grace of many third world countries which have little else in the way of resources. One bizarre artifact of our desire to exchange cash for a glimpse of "real" nature is the production of promotional videos that depict all aspects of animal life, especially the hunting and killing of prey—what nineteenth-century philosopher Herbert Spencer might refer to as pictures of "nature, red in tooth and claw." Many of these videos advertise that their footage includes lionesses hunting down some poor grazing mammal.

The hunting behavior of lions in the Serengeti National Park has been studied almost continuously since the pioneering work of George Schaller in the 1960s.[9] Researchers line up to continue Schaller's work. Among the current bearers of this torch are Craig Packer and his former students David Scheel and Robert Heinsohn,[10] whose work has uncovered many fascinating insights into various types of cooperative and noncooperative behaviors among lions.[11] Historically, lion hunting behavior, though rarely studied in detail, has been held up as a classic case of cooperation; in other words, it sure *looked* like lions were exchanging cooperative acts.

Scheel and Packer set out to examine cooperative and non-cooperative hunting behavior in lionesses in a more controlled manner. Rather than relying on anecdotes of cooperation in lions, they set out to test four specific predictions emerging from theoretical work published by Packer and Lore Rutton:[12]

1. The tendency to participate in hunts will increase with the difficulty associated with hunting a particular prey.
2. Poor hunters should hunt less.
3. Individuals should be less likely to join large hunting groups.
4. Cooperation should be more likely among kin than non-kin.

What they found was that lions typically hunt in cooperative groups when stalking large prey but hunt alone when going after small prey.[13] When the environment is "harsh" in that you can't get food alone (i.e., large prey are present), you cooperate, but when in a "mild" environment (with small prey that you can easily take down yourself), you go it alone. Further indirect support for the idea of by-product mutualism in lion hunting is evident in that lions are not more likely to hunt when relatives are in the area (no kin selection), and no evidence for reciprocity in hunting has been uncovered.

Life as a lion is not quite as simple as the "cooperate for large prey, but hunt alone for small items" rule might imply. In fact, lions, like most animals, need to decide whether to be cooperative in many different contexts, and thus things often get too complicated. One dangerous yet potentially cooperative endeavor is defending a territory against rival groups. Not all that many things besides man can hurt a lion—but other lions can and often do, particularly when fighting over territory. Given all of the benefits associated with owning a good territory, it is not surprising that lions are quite aggressive about who gets one and who doesn't. Heinsohn and Packer have shown that cooperative territory defense in female lions is quite complex.[14] Rather than simply finding cooperators (those who aid in group territorial defense) and cheaters (those who don't), lions tend to fall into four categories. Some individuals always lead when approaching a potential danger, others consistently lag behind leaders, some join in only when they are most needed, and, strangely enough, some lionesses only cooperate when they are least needed.

The social life of a lion is obviously complicated when it comes to cooperation. Yet at least we know that for lions who opt to join in a hunt, the hunting groups act in a coordinated, cooperative fashion. Things are fuzzier when it comes to group hunting in chimps.

Real Cooperators Don't Eat Bananas

Few animal species and fruits have come to be as closely linked as chimpanzees and bananas. In fact, while chimps do seem to have a strong liking for bananas, in nature they like something a bit bloodier once in a while. Given the right conditions, chimps even hunt cooperatively to get a little meat in their otherwise vegetarian diets—or so it appears. In 1978 Curt Busse found that about 1–3 percent of a chimp's diet at the Gombe National Park was meat—usually another primate, the red colobus monkey (a small tree-dwelling species weighing, on average, less than ten pounds).[15] But somewhat surprisingly, Busse found that the average amount of meat obtained per individual was no higher when they hunted monkeys in pairs than when they hunted them alone. Gombe chimps just hunt when they happen to be in groups, rather than truly undertaking cooperative group hunting. This last sentence might seem rather confusing. To better understand what Busse's findings mean, we need to examine the most comprehensive study of chimp hunting undertaken to date.

Christophe and Hedwige Boesch have been examining the chimps of Gombe and the Tai National Park (Ivory Coast) for the better part of two decades.[16] Based on early observational work on the behavior of chimps in these different populations, the Boesches knew that the Tai population undertook cooperative group hunts more often than the Gombe population. This is probably directly related to the boomerang effect described earlier in this chapter. In the Tai population, chimps received more food per individual when they hunted cooperatively in large groups. Coordinated group hunts paid off to such an extent that individuals who failed to take part paid an immediate cost. There are likely many reasons that group hunting is so profitable, the ability to get large prey and more of it being two obvious possibilities. Tai chimps actively enforce the higher payoffs associated with group hunting with a subtle, complex system of rules of access to freshly killed prey; those who don't hunt have lower probabilities of enjoying the spoils. This is precisely what we would expect if by-product mutualism is driving cooperation in the Tai population.

The story of the Gombe chimps is a bit more confusing. After all, we saw earlier that chimps may do better hunting alone than in groups at Gombe. Despite this, 52 percent of all hunts were conducted in groups. Why should this be? If you pay a cost for hunting in groups (less food per individual), why not just hunt alone? The answer is that not all group hunting is coordinated and cooperative. Chimps may physically be part of a group that is involved in hunting, just not cooperative hunting. Chris Boesch describes this phenomenon:

> A precise look at the behavior of hunters may provide an explanation. Gombe chimpanzees, when hunting in groups, start to hunt on the same group of prey, but as a rule, each follows a different target prey . . . and they do not coordinate their movements. Thus, group hunting at Gombe is better described as simultaneous solitary hunts than true cooperation.[17]

Add to this the observation that the subtle system that bars nonhunters from sharing in captured prey is absent in Gombe, and we see that cooperation is indeed a misnomer for group hunting at Gombe. Solo hunting (even when disguised as a group activity) pays, so Gombe chimps do just that.

The Wrath of Wrasse

It may not come as a great revelation to you that lions and chimps often hunt cooperatively; on the grand scale of things, these species seem to have a relatively complicated social and behavioral repertoire. But fish? Cooperatively hunting?

In fact, cooperative foraging via by-product mutualism has been found in a number of species of wrasse—beautiful tropical coral reef fish, found in the Caribbean, among other places.[18] Some of the best work on this type of cooperation is that of Susan Foster, who examined group foraging patterns in a species of wrasse called *Thalassoma lucanum,* a native of the coral reefs of the Bay of Panama.[19] Foster observed that during certain parts of the year wrasse were found alone or in small groups, while at other times large shoals were seen feeding together on the eggs of a related species. Why were wrasse cooperative foragers at some times but relative loners at others? By-product mutualism's harsh versus mild environment approach appears to provide the answer. In the case of wrasse, the harshness of their environment is directly related to the territorial defense patterns of another species, the sergeant-major damselfish.

Sergeant-major damselfish can lay more than 250,000 eggs in a single nest—a nice little meal for a group of hungry wrasse, who find such eggs quite appetizing. Naturally, the damselfish vigorously defend their nests and territories against hungry wrasse. Damselfish are rather good at this—a single wrasse has virtually no chance of getting near a nest full of edible damselfish eggs. In fact, damselfish can repel groups of up to thirty wrasse. But throw together a few hundred wrasse and the damselfish's valiant attempt to defend house and home is wasted and its eggs are eaten by the invading marauders. Make the environment for wrasse harsh (i.e., it is hard to get a good meal) and natural selection favors cooperative group hunting. Wrasse in big groups still eat more eggs per fish than going it alone (when they receive none).

But, just maybe, like the chimpanzees at Gombe, the wrasse only appear to hunt cooperatively. Perhaps they typically swim around in big groups, even when there are no damselfish eggs to hunt. If so, the burden of proof for cooperative group foraging would be higher. It turns out, though, that when damselfish are not breeding, wrasse are found in much smaller groups. Remove the harsh environment and cooperative foraging groups disappear as well.[20]

Despite the evidence for cooperation in hunting groups of wrasse, we don't really know if they *coordinate* their hunting in a manner similar to, say, lions or chimps. Cooperation in group foraging certainly pays off for wrasse, but whether they do anything as a group that exceeds the sum of their own individual actions is not clear. For that sort of evidence we need to turn to cooperative hunting in another species of fish, the yellowtail. Anecdotal evidence of even more complex cooperative foraging exists for this species.[21] In a study of group hunting by yellowtails, R. J. Schmitt and S. Strand emphasize this distinction and separate "cooperative foraging from less complex forms of group hunting behaviors by two criteria: 1) individual predators adopt different, mutually complementary roles during foraging ventures (i.e. there exists a 'division of labor'), and 2) individuals exercise 'temporary constraint' by not feeding until prey have been rendered more vulnerable."[22]

Schmitt and Strand found some evidence for both criteria. While hunting various species of prey, individual yellowtails take on unique roles that only make sense in the context of coordinated group hunting. Some yellowtails have the job of splitting a group of prey into smaller subgroups, which then are herded together by other yellowtail individuals—a remarkable division of labor by anyone's standards. Even more impressive is the finding that yellowtails have a suite of complicated group hunting maneuvers, and use different maneuvers when hunting different prey types. There is more than one way to skin the proverbial cat, and yellowtails have mastered them all.

Come and Get It (Sparrow Style)

It's one thing for animals to cooperate with each other to hunt down something no one could catch alone. It is quite a different matter to find food yourself and then scream out that you have gotten lucky and everyone is welcome to share in the festivities. How this sort of advertisement behavior evolved has been a mystery for some time. Mark Elgar, an Australian behavioral ecologist, has studied this phenomenon among sparrows in some detail and has apparently discovered the critical features that allow such cooperative advertisements to be favored by natural selection.[23]

I am sure that many at Cambridge University would have us believe that that institution is so immersed in cutting-edge research that one need only look out the window to see great science in the making. The common sparrows that Elgar studied support this claim. Watching these birds on the roof of the zoology building there, he found that "pioneer" sparrows—the first to discover the bread that he had placed on the roof—did, in fact, produce an advertisement call. Once a lucky food-finder emitted one of these "chirrup" calls, others were quick to arrive.[24] The more sparrows there were calling, the faster other sparrows arrived.

Sparrows, however, did not inevitably invite their flockmates to join in when food was found. The factor that determined whether chirrups filled the air was the size of the food item. When a piece of bread that was too large to carry off or eat by oneself was happened upon, sparrows called; otherwise they did not. Assuming that sparrows are not flying around with their kin, however, why should they call others over if the food item is large? Why not eat what you can and be done with it? Why give your competitors a free meal? It seems that sparrows do this not out of any concern with their group's success but rather because having other birds around reduces the pioneer's chances of getting attacked by a predator while eating. Sparrows that find small chunks of food would also benefit from having other birds around while they eat, but apparently a piece of bread has to be respectably large to provide the pioneer bird enough in resources to make up for the costs of having others join in the meal. Sparrows provide cooperative chirrup calls only when the summed benefits of food and antipredator behavior outweigh the costs.

The rooftops of Cambridge University are certainly a more controlled environment than the open lands of the Serengeti, but they are not a real laboratory. As such, it is still difficult to completely rule out alternative explanations for cooperative food calls. For example, is it possible that sparrows took turns chirruping, and that reciprocity played a role in this system? While it is certainly possible, in principle, to mark sparrows and try to address this question outdoors, this sort of problem is most readily attacked in a laboratory—and that is just where Kevin Clements and Dave Stephens started to examine it, using bluejays (which adjust well to the lab, at least by bird standards).[25]

Clements and Stephens tested three pairs of bluejays who were trained to feed in "Skinner boxes" (named after the famous behavioral psychologist). Each bluejay learned that certain keys in the box it lived in were associated with specific amounts of food (for example, one food pellet), given that another bird chose to peck a particular key in its own box. Bluejays learn quickly in such scenarios, and since they got to play this key game about two hundred times a day, they certainly had ample time to figure things out.

To examine whether reciprocity or by-product mutualism best explained bluejay feeding behavior, Clements and Stephens rigged the game. To start with, two bluejays were trapped in a Prisoner's Dilemma game, then the number of food pellets associated with the keys was shifted so that both bluejays always benefited from hitting a certain key (i.e., by-product mutualism payoffs); then to make life even more complicated for the birds, the game was switched back to a

Prisoner's Dilemma. The results were remarkably straightforward. When jays were in a Prisoner's Dilemma, they never cooperated, but when the payoffs simulated the "harsh" environment of by-product mutualism theory, the birds were very cooperative. Clements and Stephens argue that bluejays make their decision according to the here and now and give little weight to future payoffs. The future is the key to the Prisoner's Dilemma game and reciprocity, but the present is what counts in the world of by-product mutualism.

So far, our examples of by-product mutualism and harsh environments have focused exclusively on obtaining food, but the flip side of the coin—avoiding being eaten—can also favor various types of cooperation.

Birds That Sometimes Cry Wolf

Among the (declining) myriad of creatures in the Amazon forest, there are numerous multispecies flocks of birds—that is, different bird species living in a single group. Within such groups, there are "leader" species that are front and center in group foraging runs and that also give alarm calls when they sight predators like hawks. Two such leader species in the Amazon understory are the white-winged shrike-tanager and the bluish-slate antshrike. These two different birds have come to an amazing realization about power: it can be used in all sorts of deceptive ways to get what you really want.

Foraging in Amazonian mixed-species flocks usually involves eating insects that are flushed from the ground. Often when tanagers or antshrikes are in hot competition for one such insect, one of them will do something quite intriguing. It gives out a false alarm call, which then causes a cascade of events. The competitors for food (other birds in the group) head for cover, and the alarm caller gets not only the food that inspired the deception but also all the insects that are flushed out when everyone else heads for the hills in a chaotic whirlwind. Charles Munn found that 106 of 718 alarm calls he heard were false calls, in that no predator was in the vicinity and the above sequence of events usually unfolded.[26] Similar stories are told about various species of birds throughout the world.[27]

Remarkable as this story is, deceptive alarm callers are not all that smart. When giving a genuine alarm call, sentinels typically remain motionless on partly hidden perches. But, when emitting false calls, alarmist birds fly out in the open—a very dangerous thing to do, if a predator is truly in the area. Despite being intelligent enough to deceive others, they haven't really mastered the art of chicanery, for if they had, they'd not only voice a call but act the way scared birds act when danger is about. Of course, it is possible that natural selection has not favored such acting skills, since merely giving the call works so well. Yet that in many ways begs another question about cognitive complexity: why haven't the birds that keep getting bamboozled figured out that if an alarm caller doesn't head for the hills himself, then he is probably faking it? We simply don't know, nor has anyone even addressed the problem.

Although the "birds that cry wolf" story is true, it is the exception rather than the rule for alarm calling. Most of the time alarm callers are genuinely warning others, not trying to deceive them. But, given that alarm callers are not related in many species (see Chapter 1 for the case of related callers), why would any animal in its right mind emit an alarm? Why not just hide?[28]

Rauno Alatalo and Pekka Helle have been examining just such questions in willow tits *(Parus montanus)*.[29] Willow tits live in small flocks during the winter, and most members are unrelated.[30] As if life was not bleak enough for willow tits in central Finland in the winter, Alatalo and Helle exposed them to a model of a hawk flying by to examine alarm calling behavior. The most novel aspect of their findings is that dominant group members were more likely to give alarm calls than were subordinates. But wouldn't we expect just the opposite? Shouldn't dominant individuals use the power asymmetry in groups to somehow get subordinates to give dangerous alarm calls? Alatalo and Helle suggest the following resolution to this problem. Dominant males get most of the matings in a flock during a given season, so for them the benefits of alarm calling outweigh the costs. This is not the case, however, for underlings (subordinates). From a subordinate's perspective, it is probably a good thing to increase the chances that a dominant bird gets taken by the predator—so much the better for your chances of mating! So, dominants cooperate because it pays to do so and subordinates cheat (most of the time) for the exact same reason.

Now, it is one thing to give out an alarm call, but it is a completely different can of worms, so to speak, to actually attack a potential predator.

The Mob Mentality

People tend to like underdogs, especially when it comes to fights. Few people root for Goliath over David, and even though we may bet on the favorite in a prize fight, many of us, deep in our hearts, would be excited to see the downtrodden challenger level the champ. Sylvester Stallone (playing Rocky Balboa in the *Rocky* movies) made a fortune capitalizing on this desire of ours.

Mobbing behavior in animals, in which many small individuals attack and often chase away a predator, is the closest thing in the animal world to the victory of the underdog.[31] Seeing a flock of starlings attack a crow (four or five times the size of any of its attackers) is a pulse-raising event. I've seen this once myself and I couldn't help rooting for the starlings. I would have thrown a rock at the crow, if the whole drama hadn't ended within a few seconds.

Starlings are not alone in fighting such battles; mobbing has been documented in many different species of birds.[32] In addition to driving predators away, mobbing appears to have a much more subtle function: it teaches youngsters what is and what isn't a predator. Such information is obviously quite valuable to a young bird. In a fascinating set of experiments, Eberhard Curio and his colleagues found that young blackbirds learned what species to mob and what species are innocuous by simply watching what other blackbirds did when a possible predator came around.[33] Curio argues that this is an incipient form of cultural transmission of the enemy concept. Whether or not this is truly a case of information transmitted via culture is unclear and to some extent a matter of definition (and there are endless definitions of culture in both biology and anthropology[34]). Still, ask yourself this question: What if we replaced the word "blackbird" with "human"—would we speak of culture then?

In addition to avian struggles in the air, mobbing has also been documented in at least five species of fish.[35] Groups of smaller prey species go out en masse and nip at potential predators until they leave the area. In the case of both birds and fish mobbing their respective predators, by-product mutualism rears its head again. It simply wouldn't work (in most cases) if one or two starlings tried to chase away a crow. There is power in numbers, and the only way to make the area safe for oneself (and, as a by-product, for others) is to mob a predator in great numbers. Predators are not all that likely to attack mobbers (although it does happen), and even if a predator does go after a particular individual, the bigger the mob, the better the odds for any given mobber. Thus the benefits of mobbing may outweigh the costs, and the next thing you know, by-product mutualism emerges.

Guideposts

Our tendency to try to outdo each other can produce competition in the strangest situations. In the spring of 1997, after a winter that saw horrendous snow and rainstorms, North Dakota was hit by destructive floods. On top of torrential rains, the snow from the half-dozen or so blizzards that hit the state that winter was beginning to melt. Forecasters predicted that one-third to one-half of the state would be under water at one time or another. That was, no doubt, an overestimate, but the combination of rain and melting snow caused billions of dollars worth of damage and scores of deaths. Yet even under such conditions competition can occur and, believe it or not, it can even enhance cooperation.

When the waters started to rise, the first talk was of the biggest flood in thirty years. As time passed, North Dakota reporters began telling of the worst flood in their state in a century. Chilling as that was, I thought that the hyperbole about the magnitude of the disaster had reached its limit. It hadn't. As soon as the century reference hit the airwaves it was belittled. Eventually, reports arose that this was the worst flooding in *five hundred* years. And so North Dakotans won the "most awesome flood" contest. Actually, it is probably fairer to say that the media were more interested in how big the flood was than were the citizens of that state.

The 1997 flood also fostered cooperation. It was a good example of the harsh environment that selects for cooperation via by-product mutualism. People in dozens of towns, small and large alike, cooperated to try to minimize the damage. North Dakotans needed to work together to fill sandbags to try to hold back nature; if they didn't they'd each pay a high price. The stakes were so high that even college students—historically not the most cooperative lot—pitched in with others to fill sandbags around the clock. Such situations just make cooperation all the more critical and all the more likely.

Pollyanna cooperation it isn't, but by-product mutualism is probably the best explanation we have for the cooperation seen during such floods. Tough times "bring out the best in us," just as theory predicts, and "the best in us" in this case happens to also be the best for each of us. Let's take a look at a few more examples of this phenomenon in humans before getting to the tough questions—like whether by-product mutualism is truly cooperation and whether we should intentionally create harsh environments to foster cooperation.

Consider the thousands of joint business ventures that take place each year in the United States. Let's pick one hypothetical fledgling venture that involves four people—Steve, Phil, Susan, and Jane. These entrepreneurs (and friends) are opening the restaurant they've all wanted to own since they were kids, and it just so happens that each has some useful expertise. Steve is an accountant, Phil is an expert in business management, Susan has worked in service industries and knows where to get what they need (food, equipment, etc.) at a good price, and Jane is the ultimate hostess with all the social contacts critical to promoting their posh new place.

Our four owners may have known each other since they were kids, but we still have to address a question that surfaces in such ventures. Supposing they could get away with it—why wouldn't each of them cheat? Of course there are many possible answers to this question, based largely on variables that we have not elaborated on here. Suppose, however, that all you knew was what was written above. Why not cheat then? Why doesn't Phil skim a bit off the top, or Susan work in a little kickback scheme?

Perhaps they would. However, I'd venture to guess that if they did, it would be during a time when business was booming. In such "mild" environments, cheating may not boomerang back on the cheater himself and make his lot worse than it was before. If times were tough—say, at the start of the venture—Steve, Phil, Susan, and Jane would each stab themselves in the foot by cheating. If Steve (or any of our four) skimmed even just a little, that might be the difference between surviving or becoming another member of the "almost made it" business club. Harsh environments, as usual, select for cooperation via by-product mutualism. Examples of this type of cooperation in humans, of course, are not restricted to the here and now, nor even to typical life in modern Western societies.

Matt Ridley, in *The Origins of Virtue,* has put forth a fascinating case of mammoths, dart throwers, reciprocity, and by-product mutualism. There is little doubt that native North Americans hunted some large mammals to extinction.[36] The hunting device known as a dart thrower, suggests Ridley, is the key to these extinctions, but not only for obvious reasons. Archaeological evidence suggests that the dart thrower emerged right about when humans started hunting big game in large numbers. Before the dart thrower, big game presented the archetypal "public goods" problem. Since game hunters almost universally shared food from their hunt,[37] why not just let someone else risk life and limb to hunt a woolly mammoth that might kill you in the process? You would get the same food either way. Indeed, argues Ridley, the existence of this dilemma is likely part of the reason why we don't see much evidence of large-scale hunting of mammoths before about fifty thousand years ago. But then came the invention of the dart thrower and things got interesting.

First and foremost, dart throwers allowed a group of hunters the luxury of launching a projectile at a mammoth and thus reducing the odds of their own injury dramatically. That alone might account for the general rise in hunting big prey once the dart thrower came about. But there is more to Ridley's argument and it is subtle. What the dart thrower did was change the sort of cooperation needed to take down elephant-sized meals. By reducing the danger to cooperators, it eased the "public goods" problem of cheaters who freeloaded (didn't hunt, but shared in the kill). The dart thrower thus made the hunt more like the harsh environment of by-product mutualism. The harsh environment was not just the size of the prey (as we saw in the lion and chimp examples), as the invention of the dart thrower per se didn't change the size of mammoths roaming the continent. The newest element of the harsh environment that might have boomeranged against a cheater was the fellow next to him, who not only wasn't happy that the cheater skipped out on the hunt while still expecting a meal, but who also happened to be carrying a rather dangerous weapon.[38] Failure to cooperate might get the cheater a dart in his back—an immediate cost to cheating (indeed) and just the environment to favor cooperation via by-product mutualism.

Is It Cooperation?

There are those evolutionary and behavioral biologists who might look at the examples we reviewed in this chapter and argue that while they are interesting studies, they are irrelevant to the area of cooperative behavior because by-product mutualism is not cooperation in the first place.

By-product mutualism is indeed the most basic of all the four paths to cooperation. It is not difficult to see how and why harsh environments might favor cooperation and quickly penalize cheating. Cheaters simply do worse than cooperators—there is no temptation to cheat, because such cheating would boomerang against you. To argue, however, that just because there is no temptation to cheat we aren't really talking about cooperation is a stretch. The counterargument is simple: by-product mutualism requires some sort of coordinated action on the part of the players involved, and because high payoffs cannot be obtained without the presence of other individuals, by-product mutualism is a legitimate path to cooperation. The question, then, is whether there are ways we can create circumstances to make this type of cooperation more likely. For example, can we foster cooperation through the conscious manipulation of group size to create harsh environments?

Imagine that you are a business owner, you pay your workers a commission on each item they make, and you are thinking of starting to build widgets. You then undertake a study, which suggests that it takes four to six workers to build a good, marketable widget, with the average number of workers being about five. How many workers do you assign to each widget? Suppose you reason as follows: If I put five or six workers on each widget, it may cost a little more, but at least I know that I'll get lots of completed projects. A reasonable enough approach, but one that opens the door for cheating via the public goods dilemma. Some groups of workers will realize that for whatever reason (that may vary from group to group) they really need only four individuals to complete the task, as our initial study suggested might be true for some cases. Now, the temptation for each worker to sit back and relax a little while the others cooperate fully is quite strong, as the product can often still be built if only that occurs. But, as we saw in the chapter on reciprocity, such behavior often leads to everyone doing the same and results in suboptimal behavior at the group level.

But what if you happened to hear on the radio one morning the results of a study about harsh environments promoting cooperation? Perhaps you might consider putting four workers on each widget. After all, the workers will know that unless they fully cooperate, no widget (and no commission) will be produced. Not only do you tell them about the study's finding, but once the workers begin building the product, they realize themselves that it is true. The temptation to cheat is then removed. All workers know that if they want a commission, they must cooperate fully with each other or they each get nothing. Creating a harsh environment—one in which there is an immediate and severe penalty to each individual if he cheats—favors cooperation out of enlightened self-interest. On occasion, four workers will cooperate and still fail to build a widget, but such costs to the business could be incorporated into your decision-making process.

We might use the logic underlying by-product mutualism to structure all sorts of interactions to favor cooperation. Parents might even use it to teach cooperation to their children. Instead of providing a reward to a child for undertaking some task himself, it might be an interesting experiment to change the task slightly (but still maintain its general structure) and make the reward contingent on cooperation with a friend (or for that matter a stranger). Creating such an environment—a scenario that is harsh in the sense that one can't do well alone—might favor genuine cooperation. Promoting such cooperation might even teach children the beauty of pro-social behavior and get them to cooperate even when it didn't immediately pay for them to do so.

If all this talk of cooperation for your own good is making you depressed, the next chapter, on "good-for-the-group" cooperation, might boost your spirits—at least to start with.

For the Good of Others?

4

Man should be willing to accept hardships for himself in order that others may enjoy wealth; he should enjoy trouble for himself that others may enjoy happiness and well-being. This is the attribute of man. . . . He who is so cold-hearted as to think only of his own comfort, such a one will not be called a man. . . . Man is he who forgets his interest for the sake of others. His own comfort he forfeits for the well-being of all. Nay, rather, his own life must he be willing to forfeit for the life of mankind.

—'Abdu'l-Baha, *Foundations of World Unity*

If only it was that easy. If only standard evolutionary theory could be turned on its head to predict that we should behave in accordance with the Baha'i religion's notion of what humanity is all about. As we shall see, neither theory nor animal studies suggest that the world should be structured exactly according to the Baha'i view, yet a somewhat toned-down version of this view is not antithetical to evolutionary thinking either, especially if you throw in a few rules to punish social parasites.

Of course, the idea that people are willing to sacrifice much for the groups in which they live is not altogether unfamiliar. One of the most carefully studied examples is the Israeli kibbutz.[1] At its inception, the kibbutz was an agriculture-based commune that was critical to Israel's success as a fledgling nation in the late 1940s. While the general structure of the kibbutz has changed dramatically over the past fifty years,[2] I would like to focus here on the early kibbutzim (plural) for the purposes of illustration.[3]

When a person joined one of the early kibbutzim, virtually all personal property was handed over to the kibbutz. Children too were envisioned as "property" of the kibbutz, rather than belonging to a particular family. All children born to kibbutz members were raised, virtually from birth, in communal houses. School, meals, and most of everyday child life centered on the children's house.[4] Parents, of course, could come and visit during the day, but children usually slept in the children's house and were raised as though all others in that house were siblings. This is evident from a study of 2,769 marriages between second-generation kibbutzniks, in which not a single union was between individuals from the same peer group.[5] Incest avoidance rules apparently are generalized to all those in one's children's house, not just one's blood kin.

The early kibbutzim were group-based communities not only in their cooperative child care arrangements but also in the way employment opportunities were distributed. Everyone—manual laborers to professionals—spent some part of his or her time preparing meals, washing dishes, mopping floors, and so on. Virtually everyone also cycled through work that was the primary generator of kibbutz income—usually, but not always, agriculture. In fact, many people spent their time exclusively on these sets of group-based activities.

Kibbutz rules often dictated much of what outsiders might view as very personal decisions. What household goods members possessed in their homes was decided by a kibbutz governing body. When enough funds existed for each house to have an item—say a can opener—then all houses would be provided with this item. Until then, no house would have one. The same sort of decision-making process underlay much of what everyday life was like on a kibbutz. If one wanted to take a vacation, money might be allocated by the kibbutz, but only after a meeting to consider what such a potential vacation might be like compared to what others on the kibbutz might have the opportunity to experience. Even when and where one ate was a function of the kibbutz, as most meals were eaten in the communal dining hall at assigned times.[6] Clearly, the kibbutz demonstrates that small groups of people can come together and form units that perpetuate group cohesiveness, group stability, and group productivity. Although they include only about 3 percent of Israel's population today, kibbutzim are still part and parcel of Israeli society.

Kibbutzim and other such entities demonstrate how cooperative and self-sacrificial we can be within our own groups; however, the downside to such cooperation is a phenomenon known as "in-group biasing." Essentially, such biases exist when individuals are willing to cooperate with others in their group but no one else. Typically this manifests itself not only in being unwilling to help nonmembers but in being outwardly antagonistic toward them. It is frightening how easily people form in-group biases. Such biases might be expected in religious groups and political parties, but as the epigraph to this chapter shows, they don't always apply to religion, though they are inherent in political systems. In-group biasing rears its head even in the most trivial of decisions—and it does so in a way that suggests we are all too ready to regard "others" as the enemy.

Henri Tajfel set out to study in-group biasing in English teenagers in the late 1960s and early 1970s.[7] Tajfel and his colleagues tested sixty-four boys who all attended a common school. Individual boys were told that they were involved in an experiment on vision and were asked to estimate the number of dots flashed on a wall. The boys were then told that as long as they were present, the researchers involved in the experiment wanted to take advantage of having a large subject pool, and so they would be asked to take part in another test. The boys were then informed that they fell into either one of two groups—those that overestimated the number of dots in the vision experiment and those that underestimated the number of dots.

Once armed with this information, each of the teenagers in Tajfel's study was placed in a room alone and given a series of forms. The forms asked him to divide up monetary rewards and penalties between two other boys in the study, either two from the overestimating group, two from the underestimating group, or one from each group. Keep in mind that the amount of money a teenage subject received himself was not affected by how he distributed rewards and penalties to others and also that there was no face-to-face contact between individuals.

The results were striking. When a boy was asked to divide rewards and punishments between two individuals from the same group—either the one the subject belonged to or the one that he did not—then he distributed them equally. The subject, in other words, acted fairly. If, however, the choice was between someone from one's own group and an individual from the other group, subjects consistently favored members of their own group—despite having no information on the actual identity of boys in either group. Simply knowing that others overestimated dots as you did, even if you would never meet such dot overestimators, was sufficient information to cause an unequal distribution of monetary rewards and punishments. Similar results were obtained when subjects were grouped based on which of two artists they preferred.

Think for a moment about the frightening implications of

this study, should they prove to be general. We are so prepared to lump others into "same as me" or "different from me" categories and then bias our actions toward the former, that even information on nothing more than someone's tendency to over- or underestimate the number of dots on a wall can prompt a change in how we treat them. Given that in the real world groups often have legitimate counterpurposes and compete over scarce resources, in-group biasing is hardly a surprise. In fact, once we recognize the large role that between-group conflict has played in our evolutionary history, it really isn't even surprising that "us versus them" rules also manifest themselves in more benign situations like those simulated by Tajfel. Such rules have simply been so important that we immediately apply them, even in "inappropriate" contexts.

Group structure clearly has its positives and its negatives. When do the benefits outweigh the costs? Should we promote programs that favor what Matt Ridley refers to as "groupishness"?[8] One way to better understand what we should expect on balance is to examine the evolution of what is known as "group-selected" cooperation.

Natural Selection, Groups, and Superorganisms

Of the major theories examining the evolution of cooperation and altruism, none has been as contentious as that of "group selection." There have been two major sticking points. The first is whether natural selection can act at the level of the group, as well as at the level of the individual and the gene. The second point of contention centers on the possibility that groups are functionally organized, in the same sense that individuals are functionally organized. That is, just as each individual is composed of various organs united in purpose, can we argue that groups are composed of individuals functionally organized to produce group-level adaptations? To address these issues we must look back at the history of the science of evolution.

To paint Charles Darwin as anything but an advocate of the power of natural selection at the level of the individual is to distort history. Most of Darwin's ideas, as well as the examples he discussed, focused on the individual and the power of natural selection. Darwin, however, also recognized that natural selection *could* act at the level of the group and he most often raised this possibility in the realm of human cooperation. Darwin's most group-selectionist work, *The Descent of Man* (a follow-up of sorts to *The Origin of Species,* albeit thirteen years later), argues that the evolution of bravery in humans can best be understood as follows: "A tribe including many members who, from possessing in a high degree patriotism, fidelity, obedience, courage and sympathy, were always ready to aid one another, and to sacrifice themselves for the common good would be victorious over most other tribes; and this would be natural selection."[9]

One is left with little doubt that, at least with respect to humans, Darwin felt that natural selection could favor certain groups over others and that this was so even when cooperative acts had no positive effect on an individual within his own group. After Darwin, and until the early 1960s, there was a quiet coexistence between those who argued that natural selection operated primarily at the level of the individual and those who believed that natural selection at the group level was also critical to our understanding of the evolution of social behavior. In the seventy years following the publication of *The Descent of Man,* the best-known of the group selectionists (although he didn't call himself that) was Warder Clyde Allee of the University of Chicago. Allee, a pioneer in the field of animal behavior, worked on cooperative behavior in everything from worms to humans and was convinced that group living was the key to understanding such cooperation:

> We have good evidence that there are these two types of social or subsocial interactions among animals: the self centered, egoistic drives which lead to personal achievement and self-preservation and the group-centered, more-or-less altruistic drives that lead to the preservation of the group or some members of it, perhaps at the sacrifice of many others. The existence of egoistic forces in animal life has long been recognized. It is not so well known that the idea of the group-centered forces in animal life also has a respectable history.[10]

In tone this quote makes Allee sound like a true pluralist, but in practice he believed not only that group selection was more important than individual selection in the shaping of behavior, but that it was the key to promoting human cooperation. As mentioned in the Introduction, Allee's Quaker background and strong belief system probably played some role in his group selectionist views.

After Allee, the next in the line of famous and controversial group selectionists was V. C. Wynne-Edwards, who argued that group selection was *the* major force controlling population size in animals.[11] His most celebrated example of such population regulation was the daily chorusing behavior of birds. In many populations of songbirds, each morning (and sometimes evening) large numbers of birds join together and sing. Wynne-Edwards believed that such choruses were a means for birds to census their own population and then act to avoid overpopulation and possible extinction of that given group.

To those evolutionary biologists who held individual selection as paramount, Wynne-Edwards's argument was the straw that broke the camel's back. Led by such eminent biologists as George C. Williams, a strong backlash to group selection thinking emerged.[12] The core of this backlash was the individual selectionists' claim that *all* examples of group selection could be understood just as well at the level of the individual as at the level of the group, and hence invoking a more complicated entity like a group was completely unnecessary.[13]

Following G. C. Williams, Richard Dawkins took up the gauntlet and argued that not only did we not need to invoke group selection to explain the evolution of social behavior, but what really mattered were genes. For the most part, Dawkins said, individuals were merely the "vehicles" that genes used to travel across generations: "Populations may last a long while, but they are constantly blending with other populations and losing their identity. They are also subject to evolutionary changes from within. A population is not a discrete enough entity to be a unit of natural selection, not stable and unitary enough to be 'selected' in preference to another population."[14]

The work of Williams, Dawkins, and others halted the group selectionist approach toward studying social behavior, including cooperation. Not only was group selection no longer taught as a respectable approach toward the study of behavior, but it was generally vilified.[15] Just when the nails were being driven into the coffin of what has now become known as "naive" group selection, a new group selection school, led by David Sloan Wilson and Michael Wade, rose from the ashes.[16]

Wilson and Wade's view of group selection differed from that of their predecessors on many counts. To begin with, they provided detailed genetic models of how group selection might operate. Earlier naive group selection models lacked such a formal setting. In addition, while naive group selection relied on whole-scale extinction of populations as its driving force, the same was not true for "new" group selection models.

The new group selection approach to the evolution of cooperation identifies two forces: within-group selection and between-group selection. These two forces determine whether cooperation evolves in a world containing many different groups. Within any group that contains both cooperators and noncooperators, cooperators come up short. They pay the costs of cooperation, while the noncooperators pay nothing yet manage to parasitize the benefits. Within-group selection then acts against cooperation, as receiving benefits and failing to pay costs (noncooperators) always is favored over getting benefits and paying costs (cooperators). The kicker, however, is that groups with many cooperators outcompete groups with few cooperators (think back to Darwin's bravery example), and this between-group selection favors cooperation.

Whether cooperation evolves or not depends on the relative strength of both within- and between-group selection. When within-group selection predominates, cooperation is doomed; when between-group selection is strong, cooperation thrives. That is, when the benefit that cheaters receive is high and the number of cooperators in a group has little effect on between-group competition, we should see little cooperation. When cheaters receive only small benefits within groups, but groups with lots of cooperators thrive in competition with other groups, cooperation evolves. Keep in mind, though, that regardless of the conditions, group selection models never predict that everyone should cooperate, because cheaters would then thrive within groups. Similarly, one would not expect pure cheating to fare well. In such a scenario, a group with even just a few cooperators might have a big advantage over groups composed of pure cheaters. As such, group selection models produce some combination of cooperation and cheating as their end product.

Once the forces driving new group selection are understood, we are in a much better position to tackle, head on, the second issue that has played a prominent role in the debate on group selection—namely, can groups be functionally organized in the same way that individuals are? In essence, we are asking whether we might expect group selection to produce "superorganisms"—groups that have adaptations that appear uniquely designed to function at the group level and that provide an advantage when different groups compete for resources. The answer is that we should rarely, if ever, expect true superorganisms, in which individuals give up all their selfish tendencies for the welfare of the group, because group selection models typically produce both cooperators and noncooperators as their end product. If any noncooperators are present, our true superorganism is no more. We might, however, predict that something like a superorganism could be found in situations where between-group competition is high and within-group payoffs for cheating are low. Conversely, superorganism-like creatures should be rare when the payoff for cheating within one's group is high and between-group competition is weak. We shall examine one potential candidate for superorganism later in this chapter.

The debate between new group selectionists and individual selectionists rages on.[17] In order to arrive at your own understanding of this debate, it is important that you know the history we have touched on; it will allow you to view the animal examples we are about to examine in an appropriate context.

Guppy Guards and Group Selection

Let's return to the streams of Trinidad and the guppies first mentioned in the Introduction. But now let's think of these fish in the context of group selection. The streams where they live contain many large fish species that eat guppies. When a school of guppies sees a potential predator on the horizon, a bizarre event occurs. Rather than every fish running for cover, sometimes a few individuals break away from the school and move out toward the predator to obtain information about this possible menace (is it hungry? is it actually a predator? how far away is it?). Then, as if this is not strange enough, these inspectors turn around and head back for their school armed with this newly obtained information. To determine if this sequence of events is an example of group-selected cooperation we need to ask three questions: (1) Do the inspectors transfer the information to the noninspectors? (2) Is inspection dangerous for inspectors? and (3) Do groups with more inspectors do better than groups with a few inspectors? If the answer to all of these is yes, we will have laid bare the fourth path to cooperation.

To begin with, we need to know whether our cooperators (the inspectors) are providing our cheaters (the noninspectors) with a free benefit. After all, it is possible that while inspectors pay the cost of inspection, they keep the benefits to themselves instead of spreading around the information they have received. If this is the case, we need not consider whether cooperation in this system is driven by group selection, because inspectors wouldn't even necessarily be cooperators and the question would then be moot. Examining the transfer of information from inspectors to noninspectors is tricky—how can we even begin to set up an experiment to document such information cascades? The answer, according to Anne Magurran and her former student Anthony Higham, is one-way mirrors.

In addition to being wonderful props for spy movies, one-way mirrors, in combination with just the right light, can create a situation in which some fish can see a potential predator (let's call them "experienced" fish), while other fish cannot (let's refer to this group as "naive"). Once this setup is in place, we can then compare naive fish before and after they have interactions with experienced fish that could see and inspect a dangerous situation. What Magurran and Higham found was that naive fish displayed antipredator behavior only after interacting with experienced fish. That is, even though the naive fish themselves had no direct information about the predator, they acted as though a predator was in the vicinity after interacting with experienced individuals.[18] Somehow the experienced fish transferred information about danger to the naive group! Inspectors clearly pay the costs of inspection and receive the benefits, while noninspectors reap the benefits for free—they are indeed cooperators and cheaters, respectively.

I have been saying such things as "inspectors clearly pay the costs of inspection," but is there are any evidence for that? Is it possible that undertaking an inspection sortie is not dangerous? Further, is it possible that by "notifying" a predator that it has been spotted and that attempts to surprise the prey are doomed to failure, inspection is actually safer than not inspecting? This is an important question in the context of group selection, because in models of this phenomenon cooperators are depicted not only as helping others (as shown above) but also as taking on some risk. Do inspectors, in fact, take on any such risks?

To tackle the question of how dangerous inspection is, I went out and purchased ten kiddie wading pools, one meter in diameter. Into each of these pools I placed six male guppies that I could individually recognize from their color patterns, a single sunfish predator, and some floating rope that guppies could use as a safe refuge.[19] Each group of six guppies contained two males that inspected often, two that inspected occasionally, and two that rarely took on approaching a predator. The experiment itself was ridiculously simple. After thirty-six and sixty hours, I simply caught all living guppies and looked at how the tendency to survive mapped onto their inspection habits. The results were unambiguous: after sixty hours, 40 percent of the fish that rarely inspected were alive, 15 percent of those that inspected moderately were still alive, but not a single fish from those that inspected often remained.

The more likely you are to inspect, the more likely you are to be eaten. How could that be? If inspection is so dangerous, how could it evolve? Wouldn't we expect that over time natural selection would simply favor those that did not inspect over those that did? Not necessarily, if group selection is operating. Remember that in group selection models, selection works against cooperators within groups. Therefore, we might expect that when we create schools of fish where individuals differ in their tendency to inspect, the boldest inspectors in each group should be less likely to survive than their not-so-bold schoolmates, as I found.[20] Of course, what allows cooperation to evolve in these models is that between-group selection favors it—the more inspectors in a group, the better that group fares. Is there any evidence for this in guppies? A bit.

In an attempt to gauge the costs and benefits of predator inspection in the guppy, Jean-Guy Godin and I undertook some fieldwork in the Arima River in Trinidad.[21] In a slow-moving part of this river, we built a small wire mesh enclosure, within which we studied both the antipredator and feeding behavior of guppies in small groups (six individuals per group). A realistically painted model predator was moved across the mesh enclosure to simulate danger (guppies respond to this as if it were a live predator). What we found was that members of a group in which no guppy inspected displayed the highest rates of food acquisition. Now, at first glance that might suggest that groups with no inspectors (i.e., no cooperators) are at an advantage, which is exactly the opposite of what group selection thinking would lead us to expect. But it is not that straightforward, because while taking in food is usually a positive, experimental work has demonstrated that when a predator is in the vicinity, feeding makes an individual fish more likely to be eaten. You can't do everything at one time (at least not well), and when foraging, animals often let down their guard and become very susceptible to attacks.[22] Our research suggests that groups with no inspectors do very poorly in terms of survival when compared to groups containing at least one bold individual. Between-group selection might very well favor inspection—thus offsetting the within-group cost paid for this behavior—and allow inspection to evolve.

Overall, many aspects of predator inspection behavior in fish seem best explained by group selection. Inspectors pay the significant costs associated with inspection, but the benefits associated with this dangerous act are shared by all. Inspectors are at a disadvantage when compared with noninspectors in their group. Yet, at the same time, groups with inspectors do better than groups with no inspectors. The analogy with guard duty in the army is striking (not with respect to motivation, but rather to costs and benefits). When a group of soldiers are on point, those that discover a potential danger and go out to gather information are certainly in greater peril than the others in their patrol who stay back. On the other hand, if no one ventures out to investigate what might be the enemy, the entire group will be vulnerable to attack. As with the guppies, selection within groups favors cheating, but selection between groups promotes cooperation.

Even given the tendency of some fish to inspect predators, fish schools are not, by any stretch of the imagination, "superorganisms" designed to act as a unit. All sorts of selfish actions characterize life in a fish school. But bees—that's a different story.

Honey and the Superorganism

I shiver when I think of all the television footage I have seen of people voluntarily covering their entire bodies with honey. I'm convinced such people are not playing with a full deck, but that is not what frightens me—it's those bees. It seems almost as though the bee *colony* is a creature about to have its dinner. It's as if the bees are some giant organism that is more than just the sum of its bee parts—it's like looking at a superorganism. Of course, just because a lot of little bees look terrifying when they all gather together, and just because they look like a giant creature, that doesn't make them a superorganism in the biological sense. Technically speaking, a superorganism is "a collection of single creatures that together possess the functional organization implicit in the formal definition of organism."[23] That is, superorganisms are designed to perform particular functions, and those functions are specific to enhancing the group's performance in some form of group-versus-group competition. For example, honeybees may die when they sting someone, but colonies with such extreme altruists are certainly better protected than those without them. Honeybee defense is then best interpreted as a trait of the superorganismic bee colony.[24]

The idea that social insect colonies are best understood as superorganisms, produced as a result of group selection, can be traced back as far as Darwin and *The Origin of Species.* Today's most eloquent spokesman for this view is Thomas Seeley, a behavioral ecologist, whose book *The Wisdom of the Hive* is a delight to read.[25] Seeley has argued that honeybees provide us with a wonderful chance to see how a creature can possess attributes of both superorganism and organism simultaneously:

> A colony of honeybees, for instance, functions as an integrated whole and its members cannot survive on their own, yet individual honeybees are still physically independent and closely resemble in physiology and morphology the solitary bees from which they evolved. In a colony of honeybees two levels of biological organization—organism and superorganism—co-exist with equal prominence. The dual nature of such societies provides us with a special window on the evolution of biological organization, through which we can see how natural selection has taken thousands of organisms that were built for solitary life, and merged them into a superorganism.[26]

This is precisely what group selection thinking might suggest; after all, such models predict some mix of cooperative and noncooperative behavior, not pure cooperation. Although *all* members of a group rarely cooperate *fully* in any context, there is little doubt that aside from humans, social insects are the most likely candidates for the title of superorganism. Here we shall take a look at three particularly superorganismic honeybee colony traits: food gathering, hive temperature control, and colony defense.

Bringing Home the Pollen Honeybee foraging behavior has been studied in excruciating detail, both because these creatures are the sweethearts of insect evolutionary biologists and because of the economics of honey production.[27] Food-gathering behavior in these insects is truly wondrous—thousands of workers cover vast areas of ground (at least from a honeybee's perspective[28]) and then somehow all this information is processed in the hive and translated into a series of instructions for a whole cast of different honeybee workers. How does it all work? How is it possible for individual foragers to monitor their own intake rate, how their food changes in the environment, and the colony's needs? Part of the answer appears to be the famous honeybee "waggle dance," first described by Karl von Frisch almost fifty years ago.[29] Bees return from a bout of food gathering and then perform a bizarre "dance" that actually uses gravitational cues and angular information about the sun's position in the sky to tell their colonymates where they just encountered a food source. Furthermore, the intensity at which the dance is performed also provides information about the distance to a food source. To see this waggle dance on tape is to develop a new appreciation for the complexity of information transmitted by bees.[30] That being said, it is only a small portion of all the facts that honeybees use when maintaining sufficient levels of food in the colony.

How colony-level decisions about food intake are made is just starting to be understood. For example, in 1948, Martin

Lindauer found that when honeybees in a colony gather nectar at a high rate, individuals restrict their foraging to high-energy food sources.[31] But exactly *how* this colony-level response occurs was a mystery until 1989, when Seeley ran a series of field experiments using natural colonies of honeybees foraging on artificial feeders in upstate New York.[32] Using bees both "trained" and untrained to the location of a food source, he found that foragers recruit more nestmates to food sources when colony food resources are low and that this is accomplished by changes in foragers' dance patterns when arriving in the hive. Still, the critical question remained: how do foragers sense the change in colony-stored resources?

For a suite of reasons, all in themselves fascinating, Seeley knew that foragers do not do the obvious and "patrol" the colony for direct evidence of food reserves. Furthermore, honeybees do not smell empty food combs and assess colony reserves that way. Remarkably, what foraging bees do is to estimate colony-level food supplies by using the time it takes to unload their food to "food storer" bees. When a forager comes back with some nectar, she assesses how long it takes for food storer bees to pick up the new load. If it takes a while, then lots of others must be bringing in food, but if a food storer is readily available, food supplies must be low.[33] This approach to gauging colony reserves certainly paints the picture of a "group mind" in which each bee is analogous to a neuron in the brain of a giant superorganism—the hive. Hives with well-functioning neurons (bees) outcompete other hives with more selfish constituents.[34]

Warming Up the Place If the superorganism label is an appropriate one for the honeybee colony, then the colony needs to concern itself with the minute-to-minute problems organisms tend to all the time. For example, just as we must be constantly regulating our temperature to approximately 35 degrees C, so too might we expect the bee colony to regulate its temperature to some appropriate level. Just as we use various organs for this regulation, so should the colony use its constituent parts—bees—to do the same. Or is this pushing the limit on what we might expect from an insect colony? Is there really any indication of colony temperature regulation that can't be explained by individual bees simply regulating their own temperatures and thus producing a hive temperature? This is the type of evidence we need to muster when arguing that temperature regulation exists at the hive level.

There is some evidence suggesting that individual honeybees are capable of thermoregulating themselves to some extent,[35] but keeping the hive at a relatively constant temperature requires more than the summed action of each bee's attempt at self-thermoregulation.[36] This opens the door to fine-tune temperature regulation at the hive level, but, as with group-selected traits, also provides a chance for cheaters to stay warm for free (let others expend the energy to keep the hive comfortable). In the language of group selection, natural selection within hives favors letting others expend the energy necessary for hive thermoregulation, but natural selection between hives favors a finely tuned thermoregulating superorganism.

The comparison of hive thermoregulation with temperature control in vertebrates like ourselves can be taken to great lengths. If the honeybee colony is a superorganism, then analogies to vertebrate thermoregulation may provide some testable hypotheses. For example, Bernd Heinrich examined whether some bees may function like the sensory systems on our skin that monitor change and relay this information (via the nervous system) to our hypothalamus. "If a honeybee swarm or colony has bees on the periphery that act as receptors of temperature changes . . . then the analogy of the 'superorganism' might hold. If on the other hand bees receiving the temperature information on the periphery do not communicate the information . . . then 'superorganism' is inappropriate."[37]

Heinrich constructed experiments to test if bees at the core of the hive were in fact responding to the temperature conditions of the bees at the mantle, but found no such evidence, weakening the superorganism claim.[38] Heinrich also argues that *individual* bees behave in ways that keep their own body temperatures near 35 degrees C (the hive optimum), and that is further invalidation of the superorganism analogy. It may very well be the case that warming the hive temperature to a stable 35 degrees is simply a result of individuals behaving in such a way as to keep their temperature at 35 degrees C when in the hive. Thermoregulation, however, is very costly in terms of energy,[39] and if an individual *could* maintain an appropriate temperature just from being in a hive, the temptation to cheat would certainly be present.

Devastating Defenses If you doubt that honeybee colonies are at least reasonable contenders for the title of superorganism, just walk over and shake a nest of bees. The response will be swift, painful, and very, very organized. Antipredator behavior in honeybees is extremely complex and mindbogglingly cooperative. There are distinct "defender castes," and caste determination is genetic.[40] Colony defenders can be further subdivided into two groups: defenders (who attack very dangerous predators) and guards (who guard the nest from inter- and intraspecific parasites and may attack less dangerous threats to the nest.[41] It's as if individual honeybees were just different components of the superorganismic hive's immune system.

The Queens of Peace

You might not think that below the sands of the Sonoran Desert would be the most obvious place to find two of the most discussed examples of group-selected cooperation, but that is just where Steven Rissing, and subsequently Gregory Pollock, found them. Two ant species, *Messor pergandei* and *Acromyrmex versicolor,* capture the essentials of what a system needs for group selection to produce altruistic and cooperative behaviors.[42] Let's set the stage first and then, in turn, look at each of these cases in some detail.

Many insect species undergo cooperative colony founding, in which two or more queens initiate a new nest and cooperate in raising workers of the next generation. Most such cases, however, involve queens that are highly related to one another (often sisters). Ants, for a whole series of reasons, are the exception to the rule and queens are often unrelated.[43] Colony foundation is usually claustral—that is, early nests are formed and sustained underground, safe from the vagaries of desert weather and dangerous predators. One problem with claustral life is how to obtain food. Either there must be some internal source or someone has to go out and leave the safety of the nest. For our two examples, *Messor* uses stored body fat and avoids this problem, but *Acromyrmex* does not and faces the difficult dilemma of who goes out to bring back the food.

Measure various things going on in a *Messor pergandei* ant nest and you may find some surprises. Within a nest of typically two to five queens that are completely unrelated to each other, there is hardly a hint of any selfish behavior. Queens don't fight, they don't snoop around and try to destroy each other's eggs, or anything of the sort. Rather, they all raise about the same number of eggs and seem to share in the duties necessary to keep the colony running. Why doesn't anyone cheat? If you are around unrelated individuals, why not try to skip out on your share of the work, or perhaps even eat a few of the eggs of your nestmates (you get a meal and knock off competition for your own offspring)? Within-group selection most certainly does favor exactly that, so if this force drove our system, cooperation should be absent. It may be that strong between-group selection favors cooperation even more strongly. But does it, and if so, how is it manifested?

Cooperation within groups of *Messor* is only half the picture. Once a few broods of workers are produced in a nest, all hell breaks loose between colonies. There are often many starting nests of *Messor* near one another, but only a single good area to live in—maybe some shade under the lone tree in the vicinity. Colonies of *Messor* fight vigorously for the slot of desired land and undertake brood raids, during which they plunder and capture and carry away the young from opponent nests. Such between-group battles produce cooperation *within* groups. The more the queens cooperate within a nest, the more workers they produce as a unit and the greater their chances of winning the battles they face. Thus, even though natural selection within nests favors getting rid of offspring that might be competitors of your own brood, natural selection between nests favors being cooperative and not undertaking such actions, as nests with many workers fare well.

Two lines provide further evidence in support of this claim. Work done in another population of *Messor,* this time in Southern California,[44] demonstrates that when nests are more widely dispersed, the brood raiding found in the Sonoran

Desert is absent. What happens when the *between-group* selection component of our equation (brood raiding) is removed from the picture? Queens rarely nest together and when two queens are placed in the same nest, they fight to the death. Remove between-group selection and all that remains is within-group selection for cheating.[45] Cooperation evaporates. Furthermore, even within our population that does undertake brood raiding, once the battle is complete and only a single nest remains, queens in that nest now fight until only a single one survives. Again, remove between-group selection and cooperation goes the way of the passenger pigeon.[46]

Acromyrmex versicolor's story is even more fascinating. In this species, a single queen serves as "group forager." She alone faces the dangers inherent in leaving the nest, yet all the queens share in the food that she brings back. Just as in the *Messor* example, queens are unrelated, live peacefully, and produce about the same number of eggs. How a single queen decides (or is chosen) to be the forager remains a mystery, but it appears that this decision is not a coercive one, in that it is not forced upon a particular queen by other ants. Once a queen takes on that role, however, she retains it for life and may even be punished by her nestmates if she stops her food-gathering activities.

Now, in addition to explaining cooperation among queens in the sense of complete within-group harmony, we need to further explain how natural selection could ever favor the role of "group forager." Why not simply let someone do it and eat for free? Again, the answer lies in brood-raiding. *Acromyrmex* queens find themselves in the same predicament as *Messor* queens: severe between-group competition for that most precious of resources, a good territory. The larger the group, the greater the chances of being the victorious group; and having a single, specialized forager is apparently the most efficient means of bringing in the optimal amount of food, which translates into lots of workers.[47] In *Acromyrmex,* we see what a strong force between-group selection can be, in that it can even overcome all of the within-group benefits of a passive sort of cheating—that is, letting another queen take on the dangerous role of forager.

Guideposts

Folklore has it that one way to unite all the people on earth is to have the planet invaded by aliens from a far-off world. Group selection theory suggests that this might be sound logic. Right now, as far as we know, the number of planets housing potential invaders is zero. So, if we think of earth as one group, we have no groups competing with us. Under these circumstances, the benefits that cheaters get in a group (known as humanity) is not made up for by some between-group (or between-planet) competition that favors cooperation, and hence cooperation at the planet level fails. No invaders, no grand harmony. Barring invasion, however, where might we look for a good case of group selection on earth? And what might we learn from such a case?

Suppose that bees somehow got together and wrote a treatise on getting along with one another. I suspect they would write a document similar to the Hutterites' "Epistle on Brotherly Community."The Hutterites follow a fundamentalist religion and bear some resemblance in lifestyle to the Mennonites, the Amish, and other Anabaptists.[48] One could hardly ask for a more striking example of group-selected cooperation. In fact, the Hutterites themselves think of their groups as the human equivalent of bee colonies:

> Where there is no community there is no love. True love means growth for the whole organism, whose members are all interdependent and serve each other. That is the outward form of the inner working of the Spirit, the organism of the Body governed by Christ. We see the same thing among the bees, who all work with equal zeal gathering honey; none of them holds back anything for selfish needs. They fly hither and yon with the greatest zeal and live in community together. Not one of them keeps any property for itself. If only we did not love our property and our own will![49]

Hutterites live in small farm-like communities, primarily in Alberta and Saskatchewan, Canada. No one in these communities owns private goods, and all actions are geared toward increasing group productivity by stressing an allegiance to God rather than to man. A remarkable division of labor exists in Hutterite groups and they have been able to thrive in what many consider marginal lands. Within-group cooperation translates into extreme productivity, both at the individual and group level. Hutterites have the highest birth rate of any known human society, and they are extraordinarily efficient at transforming unoccupied areas into productive Hutterite habitations. Throughout their five-hundred-year history, laws have often been enacted specifically to stop Hutterite communities from spreading too quickly. The province of Alberta passed the Communal Property Law, which stated that no Hutterite colony could locate within forty miles of another.[50]

The birth of a new Hutterite community, when it is allowed to occur normally, is a textbook case of group-level efficiency:

> The Hutterites' passion for fairness is perhaps best illustrated by the rules that surround the fissioning process. Like a honeybee colony,

> Hutterite brotherhoods split when they attain a large size, with one half remaining at the original site and the other half moving to a new site that has been preselected and prepared. In preparation for the split, the colony is divided into two groups that are equal with respect to number, age, sex, skills and personal compatibility. The entire colony packs its belongings and one of the lists is drawn by lottery on the day of the split.[51]

Evolutionary biologists with a penchant for cooperation cannot help but be drawn to the Hutterites and like groups. In addition to promoting cooperation within groups and extreme group-level productivity, the Hutterites have an elaborate system of rules to handle violations by cheaters. Hutterites take cheaters as a serious problem—a threat to their very being:

> If a brother or a sister obstinately resists brotherly correction and helpful advice, then even these relatively small things have to be brought openly before the Church. . . . If he persists in his stubbornness and refuses to listen even to the Church, then there is only one answer to this situation, and that is to cut him off and exclude him. It is better for someone with a heart of poison to be cut off than for the Church to be brought into confusion or blemished.[52]

This is the religious solution to the free-rider/cheater problem that has plagued psychologists and evolutionary biologists interested in cooperative behavior for decades.[53]

The Hutterites are a model of group-selected cooperation in humans. They display high levels of cooperation within their own group, and avoid open aggression toward other groups. Of course, the Hutterites are competing with other groups for new unsettled land; it is just an indirect competition (who gets there first with the appropriate resources). This difference is not trivial—the Hutterites avoid bloodshed.

The argument has been put forth, however, that the Hutterites are the exception to the rule and that group-selected cooperation in humans virtually guarantees violence. It is true that group selection is driven by between-group competition. The more cooperators in a group, the better that group fares against others and that often means violence. There is also no getting around the fact that even when competition between groups is indirect, in the long run it becomes more direct. This is simply because every resource is limited, and eventually indirect competition may prove inadequate to procure what group members need. Group selection models are built on these foundations. Despite this, human group dynamics are so fundamentally different from those of other species that group selection can operate in subtle ways that avoid the consequences of violent competition.

We need to recognize that in human group selection models, a group need not be defined as a unit that remains intact, in either space or time. That is, I can be a member of many different groups throughout a given day, and none of them need be completely physically isolated from other groups. My group over breakfast is my wife and son, my group when teaching is twenty or so students, my research group may be a dozen colleagues (not necessarily at my university), my neighborhood friends group may be a dozen folks, and so on. Some individuals may be in many of these groups, some in one, and some in none. In practice this means that any given type of group (let's say my research group) might compete with others like it (for grant money), but the overlap between group memberships makes between-group conflict less likely. In other words, I am unlikely to want my research group to be very aggressive to individuals in another research group, because it is possible that someone in that other group is on my softball team. Group selection operates, it just doesn't necessarily end in conflict between specific groups. If Protestants, Jews, Muslims, and Catholics all held on to their deep-seated views, but were members of the same groups outside the context of religion, the possibility of direct conflict based on religion would surely decrease.

In fact, there is some evidence not only that this is true for humans, but that we intentionally use the above logic to defuse possible group-versus-group conflict. For example, the French court for most of its history was often a collection of relatives of the various noblemen. Such individuals were obliged to take up residence at the royal court so that noblemen could not rebel against the king without jeopardizing their own kin. Groups of noblemen might have their own ideas about how to run the country, but the royal court saw to it that a bit of mixing across groups would suspend any manifestations of between-group conflict (as well as tighten the king's hold on power).[54]

This modified version of group selection, in a slightly varied form, has been used to suggest that we abandon "tribal" thinking and government in general. Matt Ridley has made this case quite eloquently:

> One hopes that our diplomats are concluding treaties that are in our interest, rather than relying on genetic memories of hostility between groups of apes on the savanna. But they take for granted certain things about human nature that we need not, in particular our tribalism. . . . If we were truly like dolphins and lived in open societies, there would still be aggression, violence and coalition build-

> ing, but the human world would be like a water-colour painting, not a mosaic of human populations. There would not be nationalism, borders, in-groups and out-groups, warfare. These are the consequences of tribal thinking, which itself is the consequence of our evolutionary heritage as coalition-building, troop-living apes. Elephants, curiously, do not live in closed societies either. Females aggregate in groups, but groups are not competitive, hostile, territorial or fixed in membership; an individual can drift from group to group. It is intriguing to imagine ourselves like that.[55]

Despite the lures of this libertarian view of what the world might be like, I am not suggesting that it is a good thing to divorce ourselves from our "tribal" feelings (which have some very positive attributes) or national identities. The point is that there is nothing in evolutionary psychology that proves that we can't have a world in which we belong to families, softball teams, research institutions, corporations, and nations that compete but do not descend to violence in pursuit of victory.

Possibilities and Pitfalls

Nothing will ever be attempted if all possible objections must first be overcome.

—Samuel Johnson

There are obviously many solutions to the problem of how to make people nicer to one another. Everything from religion to psychology provides parts of the answer to this most important of questions. In this book, I put forth a very simple premise: there is potentially a very powerful tool for promoting human cooperation that has been all but ignored. The literature on evolution and animal cooperative behavior is an untapped resource that might hold much in the way of useful information on human cooperation.

How to tap this potential gold mine of information to promote human pro-social behavior was outlined in previous chapters that examined each of four evolutionary paths to cooperation. We have seen ground squirrels endanger themselves by giving alarm calls to help relatives, queen ants volunteer to be the sole food gatherer when living with a bunch of individuals they don't even know, and vampire bats regurgitate blood meals to their hungry nestmates. We have journeyed into beehives and naked mole-rat colonies to see a single female producing all the offspring for an entire group and a vast array of others who seem intent on cooperating with each other to help such a queen. Fish and worms switch sexes, dividing up reproduction in a cooperative manner. Impalas cooperate with each other by pulling parasites off the hard-to-reach places on the bodies of their friends to make them more hygienic. Mongooses take turns baby-sitting for each other.

What, however, do all these examples tell us about humans? My hypothesis is that animals show us a stripped-down version of what behavior in a given circumstance would look like without moral will and freedom. We can take various scenarios that in the animal world might not lead to cooperation, and turn them into guides that might produce human cooperation by focusing moral will on those areas. Furthermore, we can study cases of animal cooperation and see if the underlying forces can be deployed as tools to enhance human cooperation.

To argue that we can learn something about promoting human cooperation from animal studies and evolutionary principles is not the same as claiming that we are "merely animals." At one level this is true—technically we classify ourselves as animals because of the morphological, anatomical, and biochemical similarity we share with them and do not share with plants, fungi, bacteria, or rocks. However, when biologists speak of humans as just animals, they often imply that we are not unique. But we are. Culture dictates human day-to-day activities in a way that is fundamentally different and more interesting than in other animals. Everything we do is in one way or another affected by the particular society we live in. This is so obvious that even scientists recognize it; I've heard half-serious arguments about renaming modern man *Homo culturas.* It is our cultural complexity that might allow us to take themes that work their way through many examples of animal cooperation and transform them via culture into suggestions regarding human cooperation. The plain facts of our culture allow us to use information from such animal scenarios to help us better understand our own social behavior.

As emphasized earlier, to generalize from any single example of cooperation in animals would be foolish. Just because the costs and benefits of predator inspection in guppies may be similar to guard duty in the army is no reason to suggest that guppy inspection per se can teach us anything about what the behavior of soldiers should be. But if we concentrate on underlying themes of cooperation in animals, they do provide us with some material for making preliminary suggestions about how to structure human social dynamics. If we do nothing at this point except recognize the potential importance of evolutionary approaches to animal behavior, we will have made significant strides in the right direction.

An Embarrassment of Riches

If one accepts the notion that studies of evolution and animal social behavior can be enlightening, we must come up with criteria for determining which of these studies should be given the most weight when we apply them toward promoting human cooperation. After all, there are hundreds of animal studies of cooperation; surely some must be more important than others. There are at least three ways to rank animal studies of cooperation. First, we could give highest priority to studies on cooperation in the species that most resemble us. Work on our closest evolutionary relatives—chimps, gorillas, and orangutans—would then be the focus of our attention. Not only do these species look most like us, they are also closest to us in biochemical composition and cognitive ability. This is not to say that such species are almost as smart as we are, or that they possess anything resembling morality or free will, but merely that they are cognitively much more similar to humans than are any other species. Not surprisingly, chimpanzees have received much attention from behavioral ecologists (most notably Jane Goodall and Frans de Waal[1]). All else being equal, primates seem like a logical launching pad for studying cooperation in humans.

But all else is never equal, and there are other criteria that one might adopt for ranking the importance of animal studies of cooperation. A second method gives priority to the specific cooperative behaviors that most resemble human actions. These will often be primate studies, but other orders and phyla of animals have proved, and will prove, worthy of attention on this front. Third, we might focus not on the species or the behavior but on the actual costs and benefits underlying an act of cooperation. It is these costs and benefits on which natural selection acts, and hence it is this ratio that determines whether cooperation will evolve in a given context. Without an understanding of the cost/benefit approach to behavioral studies we are stopped dead in our tracks. The costs/benefits in any paired human and animal examples may share some similarities, but differences will arise since humans adopt a more sophisticated, culture-based response than nonhumans. If not for this very important caveat, one would be forced to say that the cost/benefit perspective is the only legitimate consideration when ranking animal studies for their potential significance in promoting human cooperation.

The Road to Good Intentions Is Paved with Potholes

There are pitfalls associated with each of the four paths to cooperation. At the very least we need to acknowledge the existence of these pitfalls and recognize the unintended consequences of choosing one path to cooperation over another.

Group-selected cooperation, though perhaps the most likely path to producing extreme altruism in humans (remember the Hutterites and kibbutzniks), is driven by between-group selection. The more cooperators in a group, the better that group fares against others. This often manifests itself in conflicts between groups. Individuals within these groups are cooperative, but in general groups tend to be hostile to one another. This is often true when individuals find themselves in a single group most of their lives, but even in small tribal societies, this is not the case for humans. We form groups at work, at home, at play, when eating, and so on, and this dynamic tends to keep the peace. Consequently, we need to keep in the back of our minds that if we minimize the number of groups a given individual is in, we may increase the probability of between-group conflict.

Somewhat similarly, when natural selection favors kin cooperating with each other, it often, but not inevitably, creates situations where kin groups must battle with each other. This is because kin groups are, after all, just a special kind of group and hence face the same pitfalls associated with group selection. As with group-selected cooperation, one way to mitigate this potential problem is to recognize that placing individuals from different kin groups together in situations where they have a common goal will tend to diffuse any hostility between kin groups. Such a simple fix may have dramatic implications for human cooperation, as much human cooperation still centers on the "blood is thicker than water" principle.

By-product mutualism is in many ways the simplest and most intuitive path to cooperation—cooperate when it pays in the here and now, otherwise don't. We can surely use this information to create environments that increase the benefits of cooperation and hence the frequency with which we cooperate. But at what cost? By invoking the logic of by-product mutualism (and arguably common sense) to create situations in which cooperation pays well, are we not sending a message that the only time one needs to cooperate is when it is in one's own best interest? At times that might be the right message to send, but at times it certainly isn't. For example, if I want to know if I can trust somebody to cooperate with me over the long haul—to be my friend—what I really want to know is whether or not I am dealing with a cooperative person. If someone cooperates only when the benefits outweigh the costs, I am unlikely to think much of him or her as a potential friend. And I certainly don't want to build a relationship with someone who bails out when the going gets even a little tough. So, while by-product mutualism is a powerful force promoting cooperation, it is perhaps best used in situations where people come together for a specific purpose and are unlikely to interact with each other all that much down the road.

And then there is reciprocity, the subject of countless experiments in psychology and the favorite path of evolutionary biologists. The power of reciprocity in producing cooperation is obvious from everyday interactions. We humans are the best scorekeepers on the planet—nothing else even comes close. We keep track of every imaginable detail of how others have treated us, and we certainly know who has cooperated with us in the past and who has not. Not only do we know that, but we structure most of our future interactions with others on this knowledge (when it is available). Everything from biblical injunction to parental teaching reinforces that we should be nice to others as long as they are nice to us, *but* we shouldn't be patsies for the bad guys out there. Reciprocators are not transcendently good, they are only good to those that are good to them; they do not turn the other cheek, but strike back. There is, however, an important distinction to be made here. Reciprocal cooperation might be okay with friends, but in a spouse one might want a kind of cooperation that transcends all the paths we have covered; as long as we recognize this, reciprocity can be a powerful tool for promoting cooperation.

Inevitably, we have arrived at the point where we must consider the relationship between the ideas outlined in this book and some of the principles of religion. Over the last 140 years, the relationship between evolutionary biology and religious thought has been a tumultuous one. To many people, from the time of the *Origin of Species* up through today, Darwin's radical ideas have remained a threat to their most deeply held beliefs: that God created humans and all life on the planet and that everything one needs to lead the good life can be found in the Bible. On the flip side of the coin, there are evolutionary biologists who hold that religion stands in the path of those genuinely interested in understanding what we can about this world and its inhabitants. Many, but not all, of the issues dividing evolution and religion disappear, of course, when one realizes that science is about understanding the natural world and religion focuses on the supernatural aspects of existence.

I myself am both an evolutionary biologist and a deeply religious person who holds that Dostoyevsky was correct when he wrote that "without God, all is permitted." I'm interested in everything science can tell us about our natural world, but I see great dangers in throwing out religion and God in the search for understanding. On such questions as the age of the earth, some religious individuals and some scientists may always be at odds (with the former viewing the issue as a supernatural one, and the latter as a question firmly in the natural sciences). Nonetheless, on questions such as how to make people more cooperative, religion and science can work hand in hand to make the world a better place. There is much work devoted to cooperation in each of these areas, and an exploration of each other's approach to altruism and cooperation might yield some extraordinary fruit.

There are many religious people who will be outraged when they come upon the notion that we can learn how to make God's crowning achievement—humans—better by studying animal behavior. The notion denies that *all* we need to make people act better is the Bible and commentaries on it. If you hold that life on the planet earth has not evolved over time and that our planet was created on October 23, 4004 B.C., as calculated in the seventeenth century by Bishop James Ussher in *Annals of the Old and New Testament,* there is probably little chance that you will find this book persuasive. However, biblical literalists should note that the four paths to cooperation outlined in this book were independently discussed thousands of years ago by various religious texts. Of course these texts took a very different approach to addressing the different types of cooperation, but they addressed them nonetheless. Such concordance between religious and scientific thinking might say something powerful about our theories of cooperation.

The vast majority of religious people, however, do not take every word in the Bible literally, nor do they believe that the facts of evolution challenge the existence of God. For this larger audience, I would like to attempt to bridge the gap between the approach outlined here and those outlined in religious holy books. My understanding is that many religions adopt the view that animals have no intrinsic worth and their value is simply in relation to how they benefit man. I have grappled with this notion and I continue to do so. Most of my scientific colleagues are ecologists and environmentalists who hold that all species are inherently important, in and of themselves. Any other viewpoint, to such proponents, is "speciesism." On the other hand, most of my friends outside science may feel kindly toward animals, but they strongly object to any claim that all species are equally important. To that crowd, we are the only truly important species on the planet.

I believe that the worth of animals does indeed lie in their relationship to humans. Inherently, other species are not important. What this means is that animals have no rights. What it doesn't mean is that we can do whatever we please to nonhumans. Animals may not have rights, but because we are humans, we have an obligation to treat other life forms well, whenever possible. It isn't permissible to smash a monkey's head against the wall because you feel like it, but it is equally abhorrent to say that the killing of millions of chickens to feed people is a "holocaust," as claimed by Ingrid Newkirk of People for the Ethical Treatment of Animals (PETA).[2] Furthermore, the belief that animals have no inherent worth and no rights doesn't tell us anything about what emotions we should express toward them. You may very well feel genuine love for your pet—after all, we are an emotional species. But just because you love a pet doesn't make the recipient of that love equal to a human.

To religious individuals who hold that man is God's masterpiece, I ask you to consider the following question about work on animal cooperation. If animals are here to help us, why should it not be the case that they can help us understand how to be more cooperative? Just as molecular work on yeast may help us cure various human illnesses, so may behavioral and evolutionary work on animals help us better understand human cooperation. The beauty in this is that the animals need not even be cooperative themselves. Even cases of noncooperation can provide insight. Einstein was right: "Science without religion is lame, religion without science is blind."

Notes

INTRODUCTION: THE FOUR PATHS TO COOPERATION

1. R. Foley, *Hominid Evolution and Community Ecology* (New York: Academic Press, 1984).
2. The public radio example is just one case of a general phenomenon that economists have labeled the "free-rider" or "tragedy of the commons" problem that occurs whenever a public good is dependent on individual contributions. Garret Hardin and John Baden provide more examples of this phenomenon in *Managing the Commons* (San Francisco: W. H. Freeman, 1977).
3. J. L. Weiner and T. A. Doescher, Cooperation and expectations of cooperation, *Journal of Public Policy and Marketing* 13 (1994): 259–71.
4. This argument is laid out in full by Aristotle in *Nicomachean Ethics.*
5. Adam Smith, *The Wealth of Nations,* vol. 1, p. 13 (1776).
6. See H. A. Simon, A mechanism for social selection and successful altruism, *Science* 250 (1990): 1665–68; L. Caporael, R. Dawes, J. Orbell, and A. J. C. van de Kragt, Selfishness reexamined: Cooperation in the absence of egoistic motives, *Behavioral and Brain Sciences* 12 (1989): 683–739.
7. This analogy is laid out in R. A. Shweder, Too important to be left to rational choice, *Behavioral and Brain Sciences* 12 (1989): 720.
8. *Utility* is technically defined in economics as the amount of consumer satisfaction associated with a good; a utility function plots utility against quantity of an item. See W. H. Anderson, A. Putallaz, and W. G. Shepard, *Economics* (New York: Prentice Hall, 1983). This concept was first introduced by Jeremy Bentham and his utilitarian group and later formalized by William Stanley Jevons.
9. I do not mean to suggest that the timescale and utility problems are insurmountable or that economists have not addressed them in some detail, but rather that we must recognize them as potential pitfalls.

10. D. Botstein, S. A. Chervitz, and J. M. Cherry, Yeast as a model organism, *Science* 277 (1997): 1259.
11. D. A. Sinclair, K. Mills, and L. Guarente, Accelerated aging and nucleolar fragmentation in yeast *sgs1* mutants, *Science* 277 (1997): 1313–16.
12. For more see P. W. Sherman, Nepotism and the evolution of alarm calls, *Science* 197 (1977): 1246–53; The meaning of nepotism, *American Naturalist* 116 (1980): 604–6.
13. For a review of predator inspection and cooperation see L. A. Dugatkin, *Cooperation among Animals: An Evolutionary Perspective* (New York: Oxford University Press, 1997).
14. For a review of lion cooperative hunting and a fascinating tale of life as a behavioral researcher in Africa, see C. Packer, *Into Africa* (Chicago: University of Chicago Press, 1994).
15. S. Rissing, G. Pollock, M. Higgins, R. Hagen, and D. Smith, Foraging specialization without relatedness or dominance among cofounding ant queens, *Nature* 338 (1989): 420–22.
16. Charles Darwin, *On the Origin of Species* (London: J. Murray, 1859); *The Descent of Man and Selection in Relation to Sex* (London: J. Murray, 1872).
17. Although most evolutionary biologists have genes (for a particular behavioral trait) in mind when they speak of a means whereby offspring are likely to resemble their parents, the last few years have been a time of maturation for the idea that "culture" can just as easily serve to transmit information across generations. It might appear strange that just such a statement has taken hold in evolutionary thinking only recently. After all, isn't it strikingly obvious that parents teach their kids to be like themselves, or, more subtly, that different ideas can be instilled in individuals with varying degrees of success, again presenting a means by which information can be passed down through generations? Evolutionary biologists were indeed aware of such arguments, but evolution is a science and as such its adherents require that a reasonable theoretical framework be put forth before they accept or refuse to accept an idea. The theoretical framework (i.e., the mathematical models)

has only recently emerged for explaining how cultural selection might operate.

18. See George Romanes, *Darwin and after Darwin: An Exposition of the Darwinian Theory and a Discussion of Post-Darwin Ideas* (Chicago: Open Court Publishing, 1892). Romanes and Darwin had an extensive history of exchanging letters on everything from personal to professional issues.

19. See Marge Midley's *Beast and Man: The Roots of Human Nature* (Ithaca: Cornell University Press, 1978) for more on how many philosophers, such as Plato, Aristotle, and Kant, depicted animals as evil, not just *symbols* of evil.

20. D. Todes, *Darwin without Malthus* (Oxford: Oxford University Press, 1989).

21. W. C. Allee is another good case in point. Allee was one of the more famous behavioral scientists of the 1930s, focusing his considerable scientific and political clout on questions surrounding both cooperation and the establishment of dominance hierarchies. Allee's strict Quaker background and value system may have played some role in his belief that dominance hierarchies, wherein each individual in a group keeps all those of lesser stature in line by aggressive acts, are actually a form of cooperation. If nothing else, his background and values influenced what experiments he did and did not undertake.

22. This is far from an antiquated view in biology. One of this generation's premier evolutionary biologists, G. C. Williams, still is of this notion, as he noted in a recent introduction he wrote to Huxley's works. The Huxley quote can be found in T. Huxley, *Evolution and Ethics* (New York: Appleton, 1897), p. 51.

23. E. Mayr, *One Long Argument* (Cambridge: Harvard University Press, 1991).

24. Stephen Jay Gould, Kropotkin was no crackpot, *Natural History,* July 1988, pp. 12–21.

25. E. O. Wilson, *Sociobiology* (Cambridge: Harvard University Press, 1975).

26. R. Wright, *The Moral Animal* (New York: Vintage, 1994).

27. In addition to the pop sociobiology phenomenon, there is another sociological issue at play here. Among those who study animal social behavior, one word more than any other strikes fear in the heart, and that word is "anthropomorphism"—the tendency to project human emotions and perspectives onto animals. In fact, to some behavioral ecologists, the very notion of talking about nonhuman and human cooperation in the same book would be a breach of protocol. Hence there is a gut-level reaction among many to simply avoid linking animal and human examples. For more on this see R. Mitchell, N. Thompson, and L. Miles, eds., *Anthropomorphism, Anecdotes, and Animals* (Albany: State University of New York Press, 1997).

28. An idea first introduced by philosopher John Stuart Mill. See Wright's *The Moral Animal* for more on the history of the naturalistic fallacy and its place in evolutionary thinking.

CHAPTER 1. ALL IN THE FAMILY

1. The Bible is full of such kin-oriented problems and solutions. For example, when Abraham and Lot are in dispute, they turn to kinship to solve it (Genesis 13: 5–9).

2. To see this, let's do the calculations. First, recall that all of the genes you and your brother share come from one place—mom and dad. There is simply no other place that they could have come from. As such, there are two ways, *and only two ways,* that your sibling and you could share a gene or genes for being cooperative—via mother or father. If your mother has a specific gene for being cooperative, genetics dictates that there is a 1 in 2 probability that she will pass it on to you and likewise a 1 in 2 probability that she will pass it on to your brother. To calculate the chances that she passes it on to *both* you and your sibling, we simply multiply these odds and find a 1 in 4 chance that *mother* is the reason that you and your brother share the gene(s) for cooperation. Similarly, there is a 1 in 4 probability that *father* is the reason that you two share the gene for cooperation (replace "mother" with "father" in the last sentence and you can prove this to yourself). This exhausts our possibilities, and so to calculate the chances that you and your sibling share a gene for cooperation through *either* mom or dad, we add the probabilities for each of them together and obtain 1/4 + 1/4, or 1/2. Such *r* values can be calculated for any set of relatives, no matter how distant (the math just gets more detailed for more distant relatives). There are many technical sources providing details on calculating values for *r*. Essentially, a pedigree chart is drawn with links between related individuals. The number of links is then put into a mathematical formula that produces values for *r*. See J. R. Krebs and N. B. Davies, eds., *An Introduction to Behavioural Ecology,* 3rd ed. (London: Blackwell Science, 1993).

3. In *The Extended Phenotype* (Oxford: Oxford University Press, 1982), Dawkins argues that individuals are merely vehicles designed by genes to make more copies of themselves.
4. See P. G. Hepper's book *Kin Recognition* (Cambridge: Cambridge University Press, 1991) for more on the varied and contentious issues surrounding if and how animals are able to recognize kin.
5. H. K. Reeve, The evolution of conspecific acceptance thresholds, *American Naturalist* 133 (1989): 407–35.
6. David Sloan Wilson and Elliott Sober touch on this question in Reintroducing group selection to the human behavioral sciences, *Behavioral and Brain Sciences* 17 (1994): 585–654.
7. W. D. Hamilton, the founder of modern kin selection theory, states his views quite clearly: "It obviously makes no difference if altruists settle with altruists because they are related (perhaps never having parted from them) or because they recognize fellow altruists as such or settle together because of [pleiotropic] effects of the gene on habitat preference." W. D. Hamilton, Innate social attitudes in man: An approach from evolutionary genetics, in *Biosocial Anthropology,* ed. R. Fox (New York: Wiley, 1975), pp. 133–55.
8. Many consider J. B. S. Haldane, Ronald Fisher, and Sewall Wright as the "founding fathers" of modern evolutionary biology. Each of these was both a biologist and a mathematician, and their theoretical work forms the baseline from which most modern ideas in evolutionary biology are derived.

9. We are limiting ourselves here to the case where a colony has a single rather than multiple queens, as this is the most common case in social insects.
10. F. L. Ratnieks and P. Visscher, Reproductive harmony via mutual policing by workers in eusocial hymenoptera, *American Naturalist* 132 (1988): 217–36; Worker policing in the honeybee, *Nature* 342 (1989): 796–97.
11. The term "policing" was first applied to insects by George Oster and E. O. Wilson in *Caste and Ecology in the Social Insects* (Princeton: Princeton University Press, 1978).
12. R. Boyd and P. Richerson, Punishment allows the evolution of cooperation (or anything else) in sizable groups, *Ethology and Sociobiology* 13 (1992): 171–95; R. Axelrod, An evolutionary approach to norms, *American Political Science Review* 80 (1986): 1101–11.
13. R. H. Crozier and P. Pamilo, *Evolution of Social Insect Colonies: Sex Allocation and Kin Selection* (Oxford: Oxford University Press, 1996).
14. For more on honeypot ants and a guide to the primary literature on this group, see Bert Holldobler and E. O. Wilson's Pulitzer Prize–winning *The Ants* (Cambridge: Harvard University Press, 1990).
15. This story is outlined in detail in the preface to P. W. Sherman, J. Jarvis, and R. Alexander, eds., *The Biology of the Naked Mole-Rat* (Princeton: Princeton University Press, 1991).
16. J. Jarvis, M. O'Riain, N. Bennett, and P. W. Sherman, Mammalian eusociality: A family affair, *Trends in Ecology and Evolution* 9 (1994): 47–51.
17. In particular, see chapters by Lacey, Lacey and Sherman, and Pepper et al. in Sherman et al., *Biology of the Naked Mole-Rat.*
18. DNA fingerprinting experiments show $r = 0.81$. H. K. Reeve, D. F. Westneat, W. A. Noon, P. W. Sherman, and C. F. Aquadro, DNA 'fingerprinting' reveals high levels of inbreeding in colonies of the eusocial naked mole rat, *Proceedings of the National Academy of Science, U.S.A.* 87 (1990): 2496–2500.
19. M. J. O'Riain, J. U. Jarvis, and C. Faulkes, A dispersive morph in the naked mole-rat, *Nature* 380 (1996): 619–21.

20. C. G. Faulkes, D. Abbott, C. Liddell, L. George, and J. Jarvis, Hormonal and behavioral aspects of reproductive suppression in female naked mole-rats, in Sherman et al., *Biology of the Naked Mole-Rat,* pp. 426–45.
21. Rood provides an excellent background reading on the demographics and behavior of the dwarf mongoose. See J. P. Rood, Dwarf mongoose helpers at the den, *Zeitschrift für Tierpsychologie* 48 (1978): 227–87; Group size, survival, reproduction and routes to breeding in dwarf mongooses, *Animal Behaviour* 39 (1990): 566–72.
22. Rood also discusses other possible cases of baby-sitting males in the banded mongoose (*Mungos mungos*) and in meerkats (*Suricata suricata*) in J. P. Rood, Banded mongoose males guard young, *Nature* 248 (1974): 176.
23. Reciprocity may also play a role in dwarf mongoose baby-sitting. Rood suggests that helpers may be reciprocated for their assistance in three different ways when those they help mature. Former recipients of help may (1) assist helpers in defending their territories, (2) form an alliance with the former helper to acquire a new territory, and (3) help to raise the young of the former helper. For more details see J. P. Rood, The social system of the dwarf mongoose, in *Recent Advances in the Study of Mammalian Behavior,* ed. J. F. Eisenberg and D. Kleinman (American Society of Mammalogists, 1983), pp. 454–88; Ecology and social evolution in the mongooses, in *Ecological Aspects of Social Evolution,* ed. D. Rubenstein and R. Wrangham (Princeton: Princeton University Press, 1986), pp. 131–52.
24. Two excellent reviews on the topic of "helpers-at-the-nest" in birds are Skutch's beautifully illustrated *Helpers at Birds' Nests* (Iowa City: University of Iowa Press, 1987)—this is the same Skutch who introduced the term fifty-two years earlier!—and Jerram Brown's *Helping and Communal Breeding in Birds* (Princeton: Princeton University Press, 1987). For additional reading, well worth the effort, see Stephen Emlen, Benefits, constraints and the evolution of the family, *Trends in Ecology and Evolution* 9 (1994): 282–85; An evolutionary theory of the family, *Proceedings of the National Academy of Sciences* 92 (1995): 8092–99. Helpers-at-the-nest have

also been found in numerous species; see Nancy Solomon and Jeffrey French, eds., *Cooperative Breeding in Mammals* (Cambridge: Cambridge University Press, 1997).

25. Brown provides details regarding his hypotheses on the evolution of "helpers-at-the-nest" in *The Evolution of Behavior* (New York: Norton, 1975).

26. Emlen outlines the basics of group living in white-fronted bee-eaters in S. T. Emlen, Altruism, kinship and reciprocity in the white-fronted bee-eater, in *Natural Selection and Social Behavior,* ed. R. D. Alexander and D. Tinkle (New York: Chiron Press, 1981), pp. 217–29.

27. S. T. Emlen and P. Wrege, The role of kinship in helping decisions among white-fronted bee-eaters, *Behavioral Ecology and Sociobiology* 23 (1988): 305–15.

28. The positive effect of helpers on their parents' reproductive success was first demonstrated in J. L. Brown, E. Brown, S. Brown, and D. Dow, Helpers: Effects of experimental removal on reproductive success, *Science* 215 (1982): 421–22.

29. Emlen and Wrege outline many of the ways that helpers increase the productivity of the nests in which they reside. See S. T. Emlen and P. Wrege, Breeding biology of white-fronted bee-eaters at Nakuru: The influence of helpers on breeder fitness, *Journal of Animal Ecology* 60 (1991): 309–26; Gender, status and family fortunes in the white-fronted bee-eater, *Nature* 367 (1994): 129–32.

30. For more on this unique example of parent-offspring conflict see S. T. Emlen and P. Wrege, Parent-offspring conflict and the recruitment of helpers among bee-eaters, *Nature* 356 (1992): 331–33.

31. Kurt Vonnegut, *Slapstick: or Lonesome, No More!* (New York: Delacorte, 1976).

32. I. Seltzer, *The Weekly Standard,* May 18, 1997.

33. From George Will's syndicated column, May 18, 1997, Abolishing the inheritance tax.

34. Sarah Bluffer-Hrdy discusses many similar cases: Care and exploitation of non-human primate infants by conspecifics other than the mother, *Advances in the Study of Behavior* 6 (1976): 101–58.

35. See Robert Wright's *The Moral Animal* (New York: Vintage, 1994) for more on evolutionary psychology.
36. Martin Daly and Margo Wilson discuss many varieties of super-aggressive behavior in their chilling book *Homicide* (New York: Aldine de Gruyter, 1988).
37. Dennis Prager discusses this issue in detail in *Think a Second Time* (New York: Regan Books, 1995).
38. Ross Perot, *Preparing Our Country for the 21st Century* (New York: Harper, 1995).
39. S. T. Emlen, Living with relatives: Lessons from avian family systems, *Ibis* 138 (1996): 87–100.
40. For example: E. Burnstein, C. Crandall, and S. Kitayama, Some neo-Darwinian decision rules for altruism: Weighing cues for inclusive fitness as a function of the biological importance of the decision, *Journal of Personality and Social Psychology* 67 (1994): 773–89.
41. B. M. Knauft, Reconsidering violence in simple human societies, *Current Anthropology* 28 (1987): 457–500. In this paper Knauft shows that in the Gebusi people of New Guinea "homicide is inversely rather than directly related to biogenetic rankings" (p. 457).
42. Daly and Wilson, *Homicide,* p. 26.
43. My conversations with a historian at the United States Center of Military History, as well as David and Mady Segal (military sociologists at the University of Maryland), suggest that it is difficult to know for sure whether this conjecture is supported by any hard evidence.
44. N. Chagnon, *The Fierce People* (New York: Holt, Rinehart and Winston, 1968).
45. R. Ford and F. McLauglin, Nepotism, *Personnel Journal,* September 1985, pp. 57–60; F. Fukuyama, *Trust: Social Virtues and the Creation of Prosperity* (New York: The Free Press, 1995).

CHAPTER 2. ONE GOOD TURN DESERVES ANOTHER

1. G. Pollock and L. A. Dugatkin, Reciprocity and the evolution of reputation, *Journal of Theoretical Biology* 159 (1992): 25–37.

2. For more on the Wason Test and its implications for evolutionary psychology see L. Cosmides, The logic of social exchange: Has natural selection shaped how we reason? *Cognition* 31 (1989): 187–276.
3. Darwinian algorithms are discussed at length in Cosmides and Tooby's 1989 pair of papers: J. Tooby and L. Cosmides, Evolutionary psychology and the generation of culture, Part I: Theoretical considerations, *Ethology and Sociobiology* 10 (1989): 29–49; L. Cosmides and J. Tooby, Evolutionary psychology and the generation of culture, Part II: Case study: A computational theory of social exchange, *Ethology and Sociobiology* 10 (1989): 51–98. Another good reference for this material is J. Barkow, L. Cosmides, and J. Tooby, eds., *The Adapted Mind: Evolutionary Psychology and the Generation of Culture* (Oxford: Oxford University Press, 1992).
4. Many psychologists had modeled cooperation before Trivers, but not from an evolutionary perspective.
5. "One human being saving another, who is not closely related and is about to drown, is an instance of altruism. Assume that the chance of the drowning man dying is one-half if no one leaps in to save him, but that the chance that his potential rescuer will drown if he leaps in to save him is much smaller, say one in twenty. Assume that the drowning man always drowns when his rescuer does and that he is always saved when the rescuer survives the rescue attempt. Were this an isolated event, it is clear that the rescuer should not bother to save the drowning man. But if the drowning man reciprocates at some future time and the survival chances are then exactly reversed, it will have been to the benefit of each participant to have risked his life for the other. Each participant will have traded a one-half chance of dying for about a one-tenth chance. Note that the benefits of reciprocity depend on the unequal cost/benefit ratio of the altruistic act, that is, the benefit of the altruistic act to the recipient is greater than the cost of the act to the performer. . . . Why should the rescued individual bother to reciprocate? Selection would seem to favor being saved from drowning without endangering oneself by reciprocating. Why not cheat? . . .

Selection will discriminate against the cheater if cheating has later adverse affects on his life which outweigh the benefits of not reciprocating. This may happen if the altruist responds to the cheating by curtailing all future possible altruistic gestures to this individual." R. L. Trivers, The evolution of reciprocal altruism, *Quarterly Review of Biology* 46 (1971): 189–226.

6. I hold this view as well; Robert Trivers will undoubtedly be regarded by generations to come as one of the founders of this field.
7. Trivers did suggest that the mathematical theory of games might be the appropriate modeling tool, but he did not use this technique himself. He did, however, present some simple population genetics equations for the evolution of reciprocity.
8. See R. Axelrod, Effective choices in the Prisoner's Dilemma, *Journal of Conflict Resolution* 24 (1980): 3–25; and More effective choices in the Prisoner's Dilemma, *Journal of Conflict Resolution* 24 (1980): 379–403.
9. Richard Dawkins tells this story in *The Selfish Gene,* 2nd ed. (Oxford: Oxford University Press, 1989).
10. J. von Neumann and O. Morgenstein, *Theory of Games and Economic Behavior* (Princeton: Princeton University Press, 1953).
11. It is not enough that individuals can expect to interact with each other many times in the future. In addition, the exact number of future interactions must be unknown, as well as large. If this is not the case, something referred to as the "backward induction" paradox occurs. This paradox refers to the fact that if you know exactly when the last encounter with someone will be, you should cheat, since you will never encounter that individual again. Both players should do this. This logic can be shown to be true for the penultimate move of the game and every move before it until the whole process of cooperation unravels.
12. Evolutionary biologists now have a penchant for creating slightly modified variants of the TfT strategy to see how they fare in computer tournaments. Such strategies as "Tit-for-Two-Tats," "Contrite Tit-for-Tat," "Observer Tit-for-Tat," and "Generous Tit-for-

Tat" can be found in the literature on the evolution of cooperation. On the flip side of the coin, more sophisticated cheating strategies such as "Roving Defector" and "Con Artist" have also been introduced. See L. A. Dugatkin, *Cooperation among Animals: An Evolutionary Perspective* (New York: Oxford University Press, 1997).

13. For example, see Dennis Prager's book *Think a Second Time* (New York: Regan Books, 1995).

14. Interestingly, many biblical scholars do not hold that "an eye for an eye" should be taken literally. One rabbi, for instance, argued that this couldn't possibly be taken literally because should a blind man put out the eye of another, there would be no way to punish him. Rather, "an eye for an eye" was a metaphor to represent equal treatment, in contrast with other ancient laws such as Hammurabi's Code that apportioned penalties based on the status of those involved in a crime. I thank Rabbi Eric Slaton for pointing this out to me.

15. P. Colgan, The motivational basis of fish behaviour, in *The Behavior of Teleost Fishes,* ed. T. Pitcher (Baltimore: Johns Hopkins University Press, 1986), pp. 23–68.

16. E. A. Fischer, Egg trading in the chalk bass, *Serranus tortugarum,* simultaneous hermaphrodite, *Zeitschrift fur Tierpsychologie* 66 (1984): 143–51.

17. Fischer warns, however, that "it could be that egg-trading is nicer or more forgiving than simple TfT." Rather than playing pure TfT, egg swappers may use something more akin to Generous TfT, in that individuals seem to reciprocate only about 80 percent of the time, yet pairs often stay together for long periods. E. Fischer, Simultaneous hermaphroditism, Tit-for-Tat, and the evolutionary stability of social systems, *Ethology and Sociobiology* 9 (1988): 119–36.

18. E. Fischer, The relationship between mating system and simultaneous hermaphroditisim in the coral reef fish *Hypoplectrus nigricans, Animal Behaviour* 28 (1980): 620–33.

19. Richard Connor has presented an interesting alternative explanation of egg swapping, based on his "pseudo-reciprocity" model

(Pseudo-reciprocity: Investing in mutualism, *Animal Behaviour* 34 [1986]: 1652–54). As opposed to true reciprocity altruism, where it always pays to cheat once benefits have been received, no such temptation exists in pseudo-reciprocity: "In the pseudo-reciprocity paradigm, an individual A performs a beneficent act for an individual B in order to increase the probability of receiving incidental benefits from B. Because the return benefits to A derive from behaviors B performs to benefit B, there is no cheating in pseudo-reciprocity." R. C. Connor, Egg-trading in simultaneous hermaphrodites: An alternative to TIT FOR TAT, *Journal of Evolutionary Biology* 5 (1992): 523–28.

20. G. Sella, Reciprocal egg trading and brood care in a hermaphroditic polychaete worm, *Animal Behaviour* 33 (1985): 938–44; Reciprocation, reproductive success and safeguards against cheating in a hermaphroditic polychaete worm, *Biological Bulletin* 175 (1988): 212–17; Evolution of biparental care in the hermaphroditic polychaete worm *Ophryotrocha diadema, Evolution* 45 (1991): 63–68.

21. C. Packer, Reciprocal altruism in *Papio anubis, Nature* 265 (1977): 441–43.

22. "All these came as allies unto the vale of Siddim . . . four kings against the five." Genesis 14: 3–9.

23. Subsequent work on coalitions in baboons by Smuts and Noe has generally confirmed Packer's findings. Bercovitch, however, found that in his nineteen-month field study of olive baboons *(Papio cyanocephalus anubis),* males who solicited coalitions were no more likely to obtain the estrous females than any other coalition member, and males who refused to join a coalition were again solicited in the future. The reasons for the discrepancy between Packer's and Bercovitch's studies are not clear, but Hemelrijk and Ek suggest that it may be tied to the presence of a clear alpha male in the latter but not the former study. See F. Bercovitch, Coalitions, cooperation and reproductive tactics among adult male baboons, *Animal Behaviour* 36 (1988): 1198–1209; C. K. Hemelrijk and A. Ek, Reciprocity and interchange of grooming and 'support' in captive chimpanzees, *Animal Behaviour* 41 (1991): 923–35; R. Noe, Lasting alliances among adult male savannah baboons, in *Primate Ontogeny, Cognition and Social Behaviour,* ed. J. Else and P. Lee (Cambridge: Cambridge University Press, 1986), pp. 381–92.

24. B. L. Hart and L. Hart, Reciprocal allogrooming in impala, *Aepyceros melampus, Animal Behaviour* 44 (1992): 1073–83.

25. M. S. Mooring and B. Hart, Reciprocal allogrooming in dam-reared and hand-reared impala fawns, *Ethology* 90 (1992): 37–51.

Reciprocal grooming in impala is unique in two aspects: (1) Impala are unlikely to be related, allowing us to rule out a major role for kinship in grooming. See M. G. Murray, Structure of association in impala, *Aepyceros melampus, Behavioral Ecology and Sociobiology* 9 (1981): 23–33; Home range, disposal and the clan system of impala, *African Journal of Ecology* 20 (1982): 253–69. (2) Dominant individuals (in male-male interactions) are not more likely to receive a disproportionate number of allogroomings, as is the case for most primate species that groom.

26. G. Wilkinson, Reciprocal food sharing in vampire bats, *Nature* 308 (1984): 181–84. Kinship was also significant in explaining long-term association patterns, but even after kinship effects were statistically removed from the costs and benefits of group living, blood sharing played a critical role.

27. Wilkinson provides evidence that one feature of social grooming in vampires is to facilitate individual recognition of cooperators and cheaters. G. S. Wilkinson, Social grooming in the common vampire bat, *Desmodus rotundus, Animal Behaviour* 34 (1986): 1880–89.

28. Wilkinson, Reciprocal food sharing.

29. Other cases of putative reciprocity in bats include work on information exchange (in the context of food calling) in the spearnosed bat and the evening bat and cluster position in pallid bats. See G. F. McCracken and J. Bradbury, Social organization and kinship in the polygynous bat *Phyllostomus hasatus, Behavioral Ecology and Sociobiology* 8 (1981): 11–34; G. S. Wilkinson, Information transfer at evening bat colonies, *Animal Behaviour* 44 (1992): 501–18; D. R.

Trune and C. N. Slobodchikoff, Position of immatures in pallid bat clusters: A case of reciprocal altruism, *Journal of Mammology* 59 (1978): 193–95.

30. M. Lombardo, Mutual restraint in tree swallows: A test of the TIT FOR TAT model of reciprocity, *Science* 227 (1985): 1363–65.

31. A control experiment was run in which two chicks were banded, rather than replaced by dead chicks.

32. This fascinating example is presented in more depth in Axelrod's *The Evolution of Cooperation* (New York: Basic Books, 1984) and Tony Ashworth's *Trench Warfare, 1914–1918: The Live and Let Live System* (New York: Holmes and Meier, 1984).

33. O. Rutter, ed., *The History of the Seventh (Services) Battalion, the Royal Sussex Regiment 1914–1919* (London: The Times Publishing Company, 1934). This quote can also be found in Axelrod, *Evolution of Cooperation.*

34. For a review as well as an evolutionary interpretation of some of this work, see L. Caporael, R. Dawes, J. Orbell, and A. van de Kragt, Selfishness reexamined: Cooperation in the absence of egoistic motives, *Behavioral and Brain Science* 12 (1989): 683–739.

35. R. Dawkins and J. R. Krebs, Animal signals: Information or manipulation? in *Behavioural Ecology,* ed. J. R. Krebs and N. B. Davies (Sunderland, MA: Sinauer Press, 1978), pp. 282–309.

CHAPTER 3. WHAT'S IN IT FOR ME?

1. For more on this see L. Caporael, R. Dawes, J. Orbell, and A. J. C. van de Kragt, Selfishness reexamined.

2. For more about *The Wager* and evolutionary biology see P. Richerson and R. Boyd, The role of evolved predispositions in cultural evolution, or, human sociobiology meets Pascal's wager, *Ethology and Sociobiology* 10 (1989): 195–219.

3. S. A. Frank, George Price's contributions to evolutionary genetics, *Journal of Theoretical Biology* 175 (1995): 373–88.

4. R. Dawkins, *The Selfish Gene,* 1st edition (Oxford: Oxford University Press, 1976); *The Extended Phenotype* (Oxford: Oxford University Press, 1982); *The Blind Watchmaker* (New York: Norton, 1987); *The Selfish Gene,* 2nd edition (Oxford: Oxford University Press, 1989).

5. M. Ridley, *The Origins of Virtue* (New York: Viking, 1996).
6. J. L. Brown, *The Evolution of Behavior* (New York: Norton, 1975); *Helping and Communal Breeding in Birds* (Princeton: Princeton University Press, 1987).
7. J. L. Brown, Cooperation—A Biologist's Dilemma, *Advances in the Study of Behavior* 13 (1983): 1–37. Richard Connor has extended Brown's ideas on by-product mutualism to cooperative interactions between species. Connor refers to his theory as "pseudo-reciprocity" as many of the cases he examines appear, at first glance, to resemble reciprocity. Pseudoreciprocity: Investing in mutualism, *Animal Behaviour* 34 (1986): 1652–54; The benefits of mutualism: A conceptual framework, *Biological Reviews* 70 (1995): 1–31.
8. This approach is detailed in L. A. Dugatkin, *Cooperation among Animals: An Evolutionary Perspective* (New York: Oxford University Press, 1997). The terms "harsh" and "mild" environment were coined by Michael Mesterton-Gibbons, a mathematical biologist at Florida State University.
9. G. B. Schaller, *The Serengeti Lion: A Study of Predator-Prey Relations* (Chicago: University of Chicago Press, 1972).
10. Packer has written a wonderfully readable account of this work as well as what conducting research in Africa is like on a day-to-day basis: C. Packer, *Into Africa* (Chicago: University of Chicago Press, 1994).
11. Behavioral ecologists have long known that cooperative hunting is common in mammals, but only in the last five years have these studies been designed to test any models for the evolution of cooperation.
12. C. Packer and L. Rutton, The evolution of cooperative hunting, *American Naturalist* 132 (1988): 159–94.
13. Technically, Scheel and Packer found three different hunting strategies: refraining (not hunting), conforming (all hunters behave similarly), and pursuing (actively hunting). Whether refrain-

ing is a form of cheating, conforming is on par with "conditional" cooperation, and pursuing equals unconditional cooperation remains to be seen. D. Scheel and C. Packer, Group hunting behaviour of lions: A search for cooperation, *Animal Behaviour* 41 (1991): 697–709.

14. R. Heinsohn and C. Packer, Who will lead and who will follow? Complex cooperative strategies in group-territorial African lions, *Science* 269 (1995): 1260–63.
15. C. Busse, Do chimps hunt cooperatively? *American Naturalist* 112 (1978): 767–70.
16. C. Boesch and H. Boesch, Hunting behavior of wild chimpanzees in the Tai National Park, *American Journal of Physical Anthropology* 78 (1989): 547–73.
17. C. Boesch, Cooperative hunting in wild chimpanzees, *Animal Behaviour* 48 (1994): 653–67.
18. This phenomenon has also been found in striped parrotfish (*Scarus croicensis*) and in various surgeonfishes.
19. S. A. Foster, Acquisition of a defended resource: A benefit of group foraging for the neotropical wrasse, *Thalassoma lucasanum, Environmental Biology of Fishes* 19 (1987): 215–22.
20. Foster sums up her work on this system by noting: "These findings imply that effective defense of a high quality resource can favor the development of gregariousness and hence of social behavior among individuals unable to gain access to the resources as solitary foragers." Foster, Acquisition of a defended resource, p. 221.
21. R. J. Schmitt and S. Strand, Cooperative foraging by yellowtail *Seriola lalandei* (Carangidae) on two species of fish prey, *Copeia* 1982: 714–17.
22. Schmitt and Strand, Cooperative foraging, p. 714.
23. Elgar begins his work on this subject by noting: "The benefits individuals may gain from locating and joining foraging flocks have been well documented by field and laboratory studies. . . . These studies have been concerned primarily with the benefits to individuals that join an established foraging flock. However, the process of flock establishment, and the conditions under which individuals

may actively attempt to establish foraging flocks have received little attention." M. Elgar, House sparrows establish foraging flocks by giving chirrup calls if the resource is divisible, *Animal Behaviour* 34 (1986): 169–74.

24. These calls were first described in J. D. Summers-Smith, *The House Sparrow* (London: Collins, 1963).

25. K. C. Clements and D. W. Stephens, Testing models of non-cooperation: Mutualism and the Prisoner's Dilemma, *Animal Behaviour* 50 (1995): 527–35.

26. C. A. Munn, Birds that 'cry wolf,' *Nature* 319 (1986): 143–45.

27. For other examples of false alarm calls see S. Matsuoka, Pseudo warning call in titmice, *Tori* 29 (1980): 87–90; A. P. Møller, False alarm calls as a means of resource usurpation in the great tit *Parus major, Ethology* 79 (1988): 25–30; A. P. Møller, Deceptive use of alarm calls by male swallows, *Hirundo rustica:* A new paternity guard, *Behavioral Ecology* 1 (1990): 1–6.

28. Alarm calling has often been depicted as the quintessential example of altruism and has been the subject of many theoretical treatments. For a review see Dugatkin, *Cooperation among Animals.*

29. R. Alatalo and P. Helle, Alarm calling by individual willow tits, *Parus montanus, Animal Behaviour* 40 (1990): 437–42.

30. J. Ekman, Coherence, composition and territories of winter social groups of the willow tit *Parus montanus* and the crested tit *P. cristatus, Ornis Scandanavia* 10 (1979): 56–68.

31. Eberhard Curio has hypothesized that mobbing behavior forces the predator to "move on." Curio, The adaptive significance of avian mobbing. I. Teleonomic hypotheses and predictions, *Zeitschrift für Tierpsychologie* 48 (1978): 175–83. Considerable evidence, in a number of taxa, supports Curio's ideas (see Dugatkin, *Cooperation among Animals,* for a review).

32. For a review of this subject see L. A. Dugatkin and J-G. Godin, Prey approaching predators: A cost-benefit perspective, *Annales Zoologici Fennici* 29 (1992): 233–52.

33. E. Curio, U. Ernest, and W. Vieth, Cultural transmission of enemy recognition: One function of mobbing, *Science* 202 (1978): 899–901.

34. R. Boyd and P. J. Richerson, *Culture and the Evolutionary Process* (Chicago: University of Chicago Press, 1985).

35. Dugatkin and Godin, Prey approaching predators.

36. See Ridley's *The Origins of Virtue* for a fascinating debunking of the myth that Native Americans have always been model ecological citizens.

37. E. Alden-Smith and B. Winterhalder, eds., *Hunter-Gatherer Foraging Strategies* (Chicago: University of Chicago Press, 1981).

38. Ridley was much more concerned with how and why hunters share food than with the difference between hunting small and large prey before and after the invention of the dart thrower. Yet he does touch on the latter, and his arguments seem to apply to this question.

CHAPTER 4. FOR THE GOOD OF OTHERS?

1. "Kibbutz" is Hebrew for "group." Much of what has been said of the kibbutz also applies to kibbutz-like settlements called moshavs.

2. Technically, the first kibbutz was founded in 1909, but the kibbutz movement is about fifty years old. For more on the modern kibbutz see E. Ben-Rafael, *Crisis and Transformation: The Kibbutz at Century's End* (Albany: State University of New York Press, 1997).

3. There is tremendous variation across kibbutz organizations, and I am only attempting to paint a broad picture of life on one of these settlements. For more on this see Harvard University's Project for Kibbutz Studies or contact any of the many Kibbutz Federations.

4. M. Spiro, *Kibbutz: Venture in Utopia* (Cambridge: Harvard University Press, 1956); E. Levy, Family life in the kibbutz, in *A New Way of Life: The Collective Settlements of Israel,* ed. G. Baratz and N. Bentwich (London: Shindler and Golomb, 1949), pp. 54–66.

5. J. Sheper, Mate selection among second generation kibbutz adolescents and adults: Incest avoidance and negative imprinting, *Archives* *of Sexual Behavior* 1 (1971): 293–307. Many, but certainly not all, modern kibbutzim have eliminated the archetypal children's houses, and those that exist are watered-down versions of the original idea.

6. The rules regarding household goods, vacations, and meals were still in place, though in milder form, at a relatively liberal kibbutz (Ein-Dor) on which I lived in the spring and summer of 1985.
7. H. Tajfel, Experiments in intergroup discrimination, *Scientific American* 223 (1970): 96–102.
8. M. Ridley, *The Origins of Virtue*.
9. C. Darwin, *The Descent of Man and Selection in Relation to Sex*, pp. 498–500.
10. W. C. Allee, Where angels fear to tread: A contribution from general sociology to human ethics, *Science* 97 (1943): 517–25, at 519.
11. V. C. Wynne-Edwards, *Animal Dispersion in Relation to Social Behavior* (Edinburgh: Oliver & Boyd, 1962).
12. G. Williams, *Adaptation and Natural Selection* (Princeton: Princeton University Press, 1966).
13. The individual selectionists' philosophy exemplifies what is known as Occam's Razor: when two explanations are available for a given phenomenon, we are obliged to choose the simpler one.
14. R. Dawkins, *The Selfish Gene*; R. Dawkins, *The Extended Phenotype*. The quotation is from *The Extended Phenotype,* p. 100.
15. E. Sober and D. S. Wilson, *Unto Others: The Evolution and Psychology of Unselfish Behavior* (Cambridge: Harvard University Press, 1998).
16. This has also been deemed "trait-group" selection, in the more technical works on the subject. See Sober and Wilson, *Unto Others.*
17. Dugatkin and Reeve show how one might attempt to bring these two camps closer to a middle ground. L. A. Dugatkin and H. K. Reeve, Behavioral ecology and 'levels of selection': Dissolving the group selection controversy, *Advances in the Study of Behaviour* 23 (1994): 101–33.
18. A. Magurran and A. Higham, Information transfer across fish shoals under predation threat, *Ethology* 78 (1988): 153–158.

19. L. A. Dugatkin, Tendency to inspect predators predicts mortality risk in the guppy, *Poecilia reticulata, Behavioral Ecology* 3 (1992): 124–28.
20. For a different view of mortality and inspection see J.-G. Godin and S. Davis, Who dares, benefits: Predator approach behaviour in the guppy *(Poecilia reticulata)* deters predator pursuit, *Proceedings of the Royal Society of London* 259 (1995): 193–200.
21. L. A. Dugatkin and J.-G. Godin, Predator inspection, shoaling and foraging under predation hazard in the Trinidadian guppy, *Poecilia reticulata, Environmental Biology of Fishes* 34 (1992): 265–76.
22. M. Milinski, Do all members of a swarm suffer the same predation? *Zeitshrift für Tierpsychologie* 45 (1977): 373–78.
23. As defined in D. S. Wilson and E. Sober, Reviving the superorganism, *Journal of Theoretical Biology* 136 (1989): 337–56.
24. This example (honeybee superorganisms) might just as easily have been discussed in the "All in the Family" chapter, as relatedness is usually quite high within colonies. There has been much discussion on the relationship between kin selection and group selection. One commonly held view is that "kin groups" are simply one type of group and that kin selection is a subset of group selection.
25. T. Seeley, *The Wisdom of the Hive* (Cambridge: Harvard University Press, 1995).
26. T. Seeley, The honeybee colony as a superorganism, *American Scientist* 77 (1989): 546–53, at 548.
27. Seeley, *The Wisdom of the Hive.*
28. P. K. Visscher and T. Seeley, Foraging strategies of honeybee colonies in a temperate forest, *Ecology* 63 (1982): 1790–1801.
29. See von Frisch for more details on this incredible behavior, and Michelsen et al. for a recent review: K. von Frisch, *The Dance Language and Orientation of Bees* (Cambridge: Harvard University Press, 1967); A. Michelsen, B. Andersen, J. Storm, W. Kirchner, and M. Lindauer. How honeybees perceive communication dances, studied by means of a mechanical model, *Behavioral Ecology and Sociobiology* 30 (1992): 143–50.
30. Besides the waggle dance, bees also undertake "round dances" and

"sickle dances." Which of these dances is performed depends on the nature of, and distance to, the food source.

31. M. Lindauer, *Communication among Social Bees* (Cambridge: Harvard University Press, 1961). This book provides an overview of Lindauer's pathbreaking work on bee communication.
32. T. D. Seeley, Social foraging in honeybees: How nectar foragers assess their colony's nutritional status, *Behavioral Ecology and Sociobiology* 24 (1989): 181–99.
33. Seeley (Social foraging) proposed that using the mathematical theory of cues, originally developed to aid economists in predicting when to reorder merchandise, will explain why "the time to unload" is such a good cue for foragers to use when assessing colony-level food reserves.
34. For more direct but technical evidence that foraging patterns may be a colony-level trait subject to natural selection for "high" and "low" foraging lines, see R. L. Hellmich, J. Kulincevic, and W. C. Rothenbluhler, Selection for low and high pollen hoarding honey bees, *Journal of Heredity* 76 (1985): 155–58; N. W. Calderone and R. Page, Effect of interactions among genetically diverse on task specialization by foraging honey bees, *Apis mellifera, Behavioral Ecology and Sociobiology* 30 (1992): 69–92; R. E. Page and M. Fondrk, The effects of colony-level selection on the social organization of honey bee (*Apis mellifera* L.) colonies: Colony-level components of pollen hoarding, *Behavioral Ecology and Sociobiology* 36 (1995): 135–44.
35. B. Heinrich, Thermoregulation by individual honeybees, in *Neurobiology and Behavior in Honeybees,* ed. R. Menzel and A. Mercer (Berlin: Springer-Verlag, 1987), pp. 102–11.
36. E. Southwick, The honey bee cluster as a homeothermic superorganism, *Comparative Biochemistry and Physiology* 75A (1983): 641–45.
37. B. Heinrich, Temperature regulation in honey bees, in *Experimental Behavioral Ecology and Sociobiology,* ed. B. Holldobler and M. Lindauer (Sunderland: Sinauer Associates, 1985), pp. 393–406, at 396.
38. B. Heinrich, The mechanisms and energetics of honeybee swarm regulation, *Journal of Experimental Biology* 85 (1981): 61–87.
39. Heinrich, Thermoregulation.

40. G. E. Robinson, R. Page, C. Strambi, and A. Strambi, Hormonal and genetic control of behavioral integration in honeybee colonies, *Science* 246 (1989): 109–12; M. D. Breed, G. E. Robinson, and R. Page, Division of labor during honey bee colony defense, *Behavioral Ecology and Sociobiology* 27 (1990): 395–401.

41. A. J. Moore, M. Breed, and M. Moore, Characterization of guard behavior in honeybees, *Apis mellifera, Animal Behaviour* 35 (1987): 1159–67.

42. For more details see S. Rissing and G. Pollock, Social interaction among pleometric queens of *Veromessor pergandei* during colony foundation, *Animal Behaviour* 34 (1986): 226–34; S. Rissing and G. Pollock, Queen aggression, pleometric advantage and brood raiding in the ant *Veromessor pergandei, Animal Behaviour* 35 (1987): 975–82; S. Rissing and G. Pollock, Pleometrosis and polygyny in ants, in *Interindividual Behavioral Variability in Social Insects,* ed. R. Jeanne (Boulder, CO: Westview Press, 1988), pp. 170–222; R. Hagen, R. Smith, and S. Rissing, Genetic relatedness among cofoundresses in two desert ant species, *Psyche* 95 (1988): 191–201; S. Rissing, G. Pollock, M. Higgins, R. Hagen, and D. Smith, Foraging specialization without relatedness or dominance among cofounding ant queens, *Nature* 338 (1989): 420–22.

43. B. Holldobler and E. O. Wilson, *The Ants*; J. Strassmann, Altruism and relatedness at colony foundation in social insects, *Trends in Ecology and Evolution* 4 (1989): 371–74.

44. R. Ryti, Geographic variation in cooperative colony foundation in *Veromessor pergandei, Pan-Pacific Entomologist* 63 (1988): 225–57.

45. One caveat is needed to the *Messor* story. Most of the detailed work done with this species has been undertaken in the lab. When David Pfennig looked for brood raiding in the field, he found no evidence of it. Pfennig believes that in *M. pergandei,* queens are forced to join *any* nest around in order to minimize predation and the probability of desiccation, and that queens inhabiting a nest allow "joiners" because of the high cost of fighting. D. W. Pfennig, Absence of joint nesting advantage in desert seed harvester ants: Evidence from a field experiment, *Animal Behaviour* 49 (1995): 567–75.

46. S. Rissing and G. Pollock, Queen aggression.
47. Even more efficient, or so it seems, than having all queens alternate in the role of forager.
48. J. Bennett, *The Hutterite Brethren* (Stanford: Stanford University Press, 1967).
49. A. Ehrenpreis, An epistle on brotherly community as the highest command of love, in *Brotherly Community: The Highest Command of Love,* ed. Friedman (Rifton: Plough Publishing, 1650/1978), pp. 9–77, at pp. 12–13.
50. Bennett, *The Hutterite Brethren.*
51. D. S. Wilson and E. Sober, Re-introducing group selection to the human behavioral sciences.
52. Ehrenpreis, An epistle on brotherly community, p. 67.
53. In fact, David Sloan Wilson and Elliott Sober (Re-introducing group selection) have gone as far as to "translate" many Hutterite statements into the language of evolutionary biology. For example, Wilson and Sober translate the passage just quoted as follows: "Not only does the selfish individual have the highest fitness within groups, but his mere presence signifies a population structure that is not conducive to altruism, causing others to quickly abandon their own altruistic strategy. Fortunately, in face-to-face groups whose members are intimately familiar with each other, it is easy to detect overt forms of selfishness and easy to expel the offenders. When 'subversion from within' can be prevented to this degree, extreme altruism, both in thought and action, becomes evolutionarily advantageous."
54. I thank David Sloan Wilson for suggesting this example.
55. Ridley, *The Origins of Virtue,* pp. 168–69.

CONCLUSION: POSSIBILITIES AND PITFALLS

1. Jane Goodall's work on the chimpanzees of Gombe has inspired a generation of behavioral researchers. Her books include *In the Shadow of Man* (co-authored with H. Lawick; Boston: Houghton-Mifflin, 1971); *The Chimpanzees of Gombe: Patterns of Behavior* (Cam-

bridge: Belknap Press, 1986); and *Through a Window: My Thirty Years with the Chimpanzees of Gombe* (Boston: Houghton-Mifflin, 1986). Frans de Waal has written three fascinating books: *Chimpanzee Politics* (Baltimore: Johns Hopkins University Press, 1982); *Peacemaking among Primates* (Cambridge: Harvard University Press, 1989); and *Good Natured* (Cambridge: Harvard University Press, 1996).

2. See Dennis Prager's *Ultimate Issues* (July 1–15, 1997) for an interview with PETA's leader.

Index

Abel and Cain, 39
Abraham, 95, 178n1
Acromyrmex versicolor, 22–23, 155, 157–58
Adoption, 66
Africa, 21–22, 54–55, 59, 61–63, 69–70, 93–97, 114–16, 190n10
Aggression: of ant queens, 157, 198n45; of baboons, 103; bees' defensive reactions, 154–55; of floater male birds, 100–101; of humans, 65–66, 69–72, 74; of lions in defending territory, 115–16; Spencer on aggression of animals, 29, 114. *See also* Hunting; Warfare
Aging, 16–17
Alarm calling, 17–20, 123–25, 165, 192n28
Alatalo, Rauno, 124–25
Alexander, Richard, 54–55
Allee, Warder Clyde, 140–41, 177n21
Altruism, 82–83, 98, 184–85n5. *See also* Group selection path to cooperation
Amazon forest, 123–24
American Enterprise Institute, 64
Animal cooperation: author's hypothesis about relationship between human cooperation and, 14–15, 36–37, 166–67; and behavioral ecology generally, 13–17; criteria for evaluating studies of, 167–68; family dynamics path to, 17–20, 42–63; four paths to generally, 17–37; group selection path to, 22–23, 139–58; Hamilton's Rule on, 43–48; Hobbes on, 1–2; Kropotkin/Wallace versus Huxley on, 29–34; objections to author's approach to, 15–17; reciprocal transactions path to, 20–21, 90–101, 103–104; and relatedness and kinship, 42–48; selfish teamwork path to, 21–22, 114–27. *See also* specific animals
Annals of the Old and New Testament (Ussher), 172
Anthropomorphism, 178n27
Ants, 1–2, 22–23, 48, 53, 155–58, 165, 197–98n45, 198n47
Antshrikes, 123
Archimedes, 54
Arima River, 148
Aristotle, 7, 90
Army, 25–26, 71, 72, 101–03, 149
Axelrod, Robert, 83–87, 101

Baboons, 92–95, 103
Baby-sitting behavior in mongooses, 58–60, 166, 181nn22–23
Backward induction paradox, 185n11
Baha'i religion, 135
Bats, 97–99, 165, 188–89nn26–29
Bay of Panama, 118–19
Bee-eaters, 61–63, 68–69
Bees, 1–2, 24, 28, 48, 50–53, 149–55, 158–59, 195n24, 196n30
Behavioral ecology: on cooperative hunting, 190n11; and family dynamics path to cooperation, 17–20; and group selection path to cooperation, 22–23; and Hamilton's Rule, 43–48; and identity by descent, 41; and kinship, 42–44; and natural selection, 27–29, 30, 34–36, 42, 82, 87, 94–95, 107–108, 139–42; and reciprocal transactions path to cooperation, 20–21; and selfish teamwork

Behavioral ecology *(cont.)*
path to cooperation, 21–22; and study of animal cooperation generally, 13–17. *See also* specific animals
Bentham, Jeremy, 175n8
Bercovitch, F., 187n23
Between-group selection, 143, 157
Bible, 12, 21, 39, 77, 89, 95, 171–73, 178n1, 186n14, 187n22
Birds: alarm calls by, 123–25; bee-eaters, 61–63, 68–69; bluejay feeding behavior, 122–23; chorusing behavior of, 141–42; as floater males, 99–101; and helpers-at-the-nest, 60–63; mixed-species flocks in Amazon as alarm callers, 123–24; mobbing behavior in, 126–27; sparrows' foraging behavior, 120–22; willow tits as alarm callers, 124–25
Black hamlets, 92
BlindWatchmaker (Dawkins), 111
Blood sharing in vampire bats, 97–99, 165, 188n26
Bluegill sunfish, 78–79
Bluejays, 122–23
Boesch, Christophe, 117–18
Boesch, Hedwige, 117–18
Boomerang effect, 113–14
Borgerhoff-Mulder, Monique, 69–70
Bravery. *See* Risk taking
Brown, Jerram, 61, 112–13, 190n7
Brown-outs, 5–6
Business: and by-product mutualism, 128–29, 131–32; and economic approaches to study of cooperation, 9–11; Japanese corporate families, 73; nepotism in, 72–73; and parceling of benefits, 106–107; self-monitoring system in, 107; teamwork in, 72–73
Busse, Curt, 116–17
By-product mutualism, 108, 111–14, 128, 131–33, 169–70, 190n7. *See also* Selfish teamwork path to cooperation
Cain and Abel, 39
Cambridge University, 121–22
Canada, 65, 159
Caporael, Linnda, 11, 12
Caribbean, 118–20
Chalk bass, 92
Cheating: benefits of, 13; and by-product mutualism, 129–32; and floater male birds, 100; and group selection theory, 143–44; Hutterites on, 160; and Prisoner's Dilemma game, 84–90, 186n12; and reciprocal cooperation, 82–90
Child abuse, 65–66
Children, 65–66, 105–106, 133, 136
Chimpanzees, 116–18, 167–68, 199n1
Chorusing behavior of birds, 141–42
Clements, Kevin, 122–23
Coalition formation, 93–95, 187n23
Colgan, Patrick, 90
Community cooperation, 68–70
Computer tournaments, 87–90, 185–86n12
Connor, Richard, 186–87n19, 190n7
Cooperation: author's hypothesis about relationship between animal and human cooperation, 36–37, 166–67; Darwin on, 28–29, 140; four paths to generally, 17–37; Hobbes versus Locke on, 1–2, 7–9; Kropotkin/Wallace versus Huxley on, 29–34; objections to author's approach to animal and human cooperation, 15–17; pitfalls associated with four paths to, 168–71. *See also* Animal cooperation; Human cooperation; and specific animals
Cornell University, 68, 78–79
Cosmides, Leda, 79–81
Cost/benefit analysis, 109, 110–11, 168
Costa Rica, 97–99
Crows, 126, 127
Cultural selection, 176–77n17
Culture, 166
Curio, Eberhard, 126, 192n31

Daly, Martin, 65–66
Damselfish, 119
Dancing of bees, 151, 196n30
Dart throwers, 129–30
Darwin, Charles, 27–29, 30, 34, 90, 140, 143, 150, 171
Darwinian algorithm, 81
Dawkins, Richard, 84, 107–108, 111, 142, 179n3
Day, Francis, 90
De Waal, Frans, 168, 199n1
Deception, 107–108
Descent of Man and Selection in Relation to Sex (Darwin), 27, 140
Diploid organisms, 48–49
DNA fingerprinting, 56–57
Doescher, Tabitha, 6
Dostoyevsky, Fyodor, 171
Drowning man dilemma, 83, 184–85n5
Dugatkin, L. A., 194n17
Dwarf mongooses, 58–60, 166, 181nn21–23

Ecological tournaments, 87–90, 185–86n12
Economic approaches to cooperation, 9–11. *See also* Business
Ecotourism, 114
Egg swapping, 91–92, 186n17, 186–87n19
Egoistic incentive theory of cooperation, 11
Einstein, Albert, 174
Ek, A., 187n23
Electricity brown-outs, 5–6
Elgar, Mark, 121, 191–92n23
Emlen, Stephen, 61–62, 68–69
England, 137–39
Environmental awareness, 67–68
Ethiopia, 55
Evening bats, 188n29
Evolution of Cooperation (Axelrod), 101
Evolutionary biology: and cultural selection, 176–77n17; founders of, 179n8; and group selection theories, 23, 139–45; and Hamilton's Rule, 43–48; and kinship, 40–44; and natural selection, 27–29, 30, 34–36, 42, 82, 87, 94–95, 107–108, 139–42; one long argument approach of, 34–35; Price's contributions to, 111; and reciprocal cooperation, 82–90; and religion, 171–74; studies on animal cooperation, 13–17. *See also* Behavioral ecology; and specific animals
Evolutionary tournaments, 87–90, 185–86n12
Expression of the Emotions in Man and Animals (Darwin), 90
Extended Phenotype (Dawkins), 111, 179n3
Extinction, 143
Eye for an eye concept, 21, 89, 186n14

Family accounting schemes, 40–48
Family dynamics path to cooperation: bee-eater example of, 60–63, 68–69; and Cain and Abel, 39; family accounting schemes, 40–48; and human cooperation, 63–75; and kinship theory, 40–48; mole-rat example of, 54–58, 165; mongoose example of, 58–60, 166; pitfalls associated with, 169; social insect examples of, 48–54; squirrel example of, 17–20, 165
Feeding behavior, 97–99, 122–23, 165. *See also* Foraging; Hunting
Firefighters, 104–105
Fischer, Eric A., 91–92, 186n17
Fish: egg swapping behavior of sea bass, 91–92, 165; emotions of, 90; foraging by wrasse, 118–19; guppies, 20–21, 24–25, 145–49, 167; hermaphroditic species of, 90–92; mobbing behavior in, 126–27; predator inspection by, 145–49, 167
Fisher, Ronald, 48, 179n8

Floater males, 99–101
Floods, 127–28
Food-gathering. *See* Foraging
Foraging, 22–23, 118–23, 151–52, 157–58, 165, 191n20, 191–92n23, 196nn33–34, 196n30, 198n45, 198n47
Ford, Robert, 72
Former Soviet Union, 6
Foster, Susan, 118–19, 191n20
France, 162
Free rider problem, 160, 175n2
Fruit flies, 16
Fundamental Law of Nature (Hobbes), 8–9

Gardens in urban areas, 26–27
God, 109, 110, 171–73
Godin, Jean-Guy, 148
Golden Rule, 77
Gombe National Park, 93–94, 116–18, 199n1
Good Samaritan, 83
Goodall, Jane, 168, 199n1
Gorillas, 167
Gould, Stephen Jay, 35
Grooming behavior, 95–97, 166, 188n25
Ground squirrels, 17–20, 165
Group selection path to cooperation: ant example of, 22–23, 155–58, 165; bee example of, 149–55; and group selection theories, 139–45, 161–63; guppy example of, 145–49, 167; and human cooperation, 135–39, 158–63; Hutterite example of, 158–61, 169; and in-group biasing, 137–39; kibbutz example of, 135–37, 169, 193nn1–6; pitfalls associated with, 169; and predator inspection by fish, 145–49, 167
Group selection theories, 23, 139–45, 161–63
Gulf War, 95
Guppies, 20–21, 24–25, 145–49, 167
Haldane, J. B. S., 40, 48, 179n8
Hamilton, W. D., 43–46, 83–87, 179n7
Hamilton's Rule, 43–48
Hammurabi's Code, 186n14
Haplodiploidy, 49–50
Harsh environment, 113–14, 115, 119, 127–31, 190n8
Hart, Benjamin, 96
Hart, Lynn, 96
Harvard University, 193n3
Hatfield/McCoy syndrome, 69–70
Heinrich, Bernd, 153–54
Heinsohn, Robert, 114, 116
Helle, Pekka, 124–25
Helpers-at-the-nest, 60–63
Hemelrijk, C. K., 187n23
Hepper, P. G., 179n4
Hermaphroditism, 90–92
Heterocephalus glaber, 55
Higham, Anthony, 146
Historia Animalium (Aristotle), 90
Hobbes, Thomas, 1–2
Homicide, 65, 70
Homicide (Daly and Wilson), 65
Homo culturas, 166
Homologs, 16–17
Honeybees. *See* Bees
Honeypot ants, 53
Human cooperation: author's hypothesis about relationship between animal cooperation and, 14–15, 36–37, 166–67; in business, 9–11, 72–73, 106–107, 128–29, 131–32; Darwin on, 28–29, 140; economic approaches to study of, 9–11; family dynamics path to, 63–75; group selection path to, 135–39, 158–63; historical perspectives on study of, 7–13; Hobbes versus Locke on, 1–2, 7–9; Hutterite example of, 158–61, 169; importance of, 2–7; and in-group biasing, 137–39; kibbutz example of, 135–37, 169, 193nn1–6; in military, 25–26, 71, 72, 73; objections to author's approach to,

15–17; among opposing groups, 74; rational man theory of, 9–12; reciprocity in, 95, 101–108; reputation for, 77–78; selfish teamwork path to, 109–12, 127–33; and social contracts, 79–81; and uncertainty of future interactions, 78–79, 185n11
Hunting: by animals, 21–22, 112–20, 190n11, 190–91n13, 193n38; by humans, 129–30
Hutterites, 158–61, 169, 198n53
Huxley, Thomas Henry, 29–30, 31, 32, 34
Hypoplectrus nigricans, 92

Identity by descent, 41
Impalas, 24, 95–97, 165–66, 188n25
Inclusive fitness theory, 45
Income tax, 67
Infanticide, 64–65
In-group biasing, 137–39
Inheritance taxes, 64
Insects, 48–54, 123–24, 149–58. *See also* specific insects
Inspection of predators, 145–49, 167
Iraq, 95
Israel, 135–37, 193–94nn1–6
Ivory Coast, 117

Japan, 73
Jarvis, Jenny, 54–55
Jevons, William Stanley, 175n8
Job (Biblical story), 12
Johnson, Samuel, 165

Kenya, 55, 61–63, 95–97
Kibbutz, 135–37, 169, 193nn1–6
Kin Recognition (Hepper), 179n4
Kin selection theory, 43–48, 50–54, 55–57, 61–62, 66, 179n7
Kinship theory, 40–48. *See also* Family dynamics path to cooperation
Krebs, John, 107–08
Kropotkin, Petr, 30, 31–34
Kuwait, 95

Lake Nakuru National Park, 61–63
Leviathan (Hobbes), 1–2, 8
Lindauer, Martin, 151–52
Lions, 21–22, 26, 36, 64–65, 114–16
Little Red Hen, 3–4
Locke, John, 7–8, 9
Lombardo, Michael, 100
Lot and Abraham, 178n1

Magurran, Anne, 146
Mammoths, 129–30
Masai Mara Reserve, 95–97
Massachusetts Institute of Technology, 16
Mayr, Ernst, 34–35
McLaughlin, Frank, 72
Meerkats, 181n22
Messor pergandei, 155–58, 197–98n45
Mesterton-Gibbons, Michael, 190n8
Mild environment, 113–14, 115, 129, 190n8
Military, 25–26, 71, 72, 73, 101–103, 149
Mill, John Stuart, 178n28
Missing element approach, 15
Mobbing behavior, 125–27, 192n31
Mole-rats, 24, 54–58, 165
Mongooses, 24, 58–60, 166, 181nn21–23
Monkeys. *See* Chimpanzees
Moral Animal (Wright), 35–36, 178n28
Morgenstein, Oscar, 84
Moshavs, 193n1
Mungos mungos, 181n22
Munn, Charles, 124
Murder. *See* Homicide
Mutual Aid (Kropotkin), 31, 33–34

Naked mole-rats, 54–58, 165
National identities, 162–63
National Public Radio, 4–5
Natural selection, 27–30, 34–36, 42, 82, 87, 94–95, 107–108, 139–42
Nepotism, 72–73
New Testament. *See* Bible

Newkirk, Ingrid, 173
"No place like home" hypothesis, 43
Noe, R., 187n23
North Dakota, 127–28
Nuclear weapons, 6
"Nudge it over the top" approach, 15

Occam's Razor, 194n13
Old Testament. *See* Bible
Olive baboons, 187n23
One long argument approach, 34–35
One Long Argument (Mayr), 34–35
Ophryocha diadema, 92
Orangutans, 167
O'Riain, M. J., 57
Origin of Species (Darwin), 27–28, 140, 150, 171
Origins of Virtue (Ridley), 111, 129–30
Oster, George, 180n11
Oxford University, 43, 84

Packer, Craig, 93, 114–16, 187n23, 190n10, 190–91n13
Pallid bats, 188n29
Papio cyanocephalus anubis, 187n23
Parceling of benefits, 106–108
Parrotfish, 191n18
Parus montanus, 124–25
Pascal, Blaise, 109
People for the Ethical Treatment of Animals (PETA), 173
Perot, Ross, 67
Persian Gulf, 95
PETA, 173
Pfennig, David, 198n45
Police officers, 71, 104–105
Policing behavior, 50, 52–53, 180n11
Politics (Aristotle), 7
Pollock, Gregory, 155
Pollyanna cooperators, 109, 128
Pop sociobiology, 35
Predators: alarm calling to warn of, 17–20, 123–25, 165, 192n28; inspection of, by fish, 145–49, 167; mobbing behavior against, 125–27, 192n31. *See also* Aggression; Hunting
Price, George, 111
Prisoner's Dilemma game, 84–90, 100, 122–23
Pseudo-reciprocity, 186–87n19, 190n7
Public goods dilemma, 130, 132
Public radio, 4–5, 6

Radio. *See* National Public Radio
Rapoport, Anatol, 88
Rational man theory of cooperation, 9–12
Ratnieks, Francis, 52–53
Reciprocal selection, 82–83, 98, 184–85n5. *See also* Group selection path to cooperation; Reciprocal transactions path to cooperation
Reciprocal transactions path to cooperation: baboon example of, 92–95, 103; egg swapping behavior of sea bass, 91–92; evolutionary biology on, 82–90; and Golden Rule, 77; guppy example of, 20–21, 24–25; hermaphroditism example of, 90–92; and human cooperation, 95, 101–108; impala example of, 95–97, 165–66; pitfalls associated with, 170–71; vampire bat example of, 97–99, 165
Reeve, H. K., 194n17
Relatedness: calculation of, 41–42, 178–79n2; definition of, 41
Religion, 171–74. *See also* Bible
Reproduction: baboon male coalition formation in, 93–95, 103; and babysitting behavior in mongooses, 58–60, 166, 181nn22–23; bee-eaters as helpers-at-the-nest, 60–63, 68–69; egg swapping in sea bass, 91–92; floater male birds, 99–101; of hermaphrodites, 90–92; of naked mole-rats, 54–58, 165; in social insect colonies, 50–54;

switching of sexes to divide up reproduction, 24, 165
Reputation for cooperation, 77–78
Ridley, Matt, 111, 129–30, 162–63, 193n38
Risk taking, 17–21, 24–26, 71–72, 73, 140, 143. *See also* Alarm calling
Rissing, Steven, 155
Rocky movies, 125
Romanes, George, 28
Rood, J. P., 59–60, 181nn21–23
Royal families, 74
Russia, 6, 31–32
Rutton, Lore, 115

Sacrificial behavior. *See* Self-sacrificial behavior
Sayings of the Fathers, 7
Schaller, George, 114
Scheel, David, 114–15, 190–91n13
Schmitt, R. J., 120
School classrooms, 105–106
Scorekeeping in social contexts, 79–81, 96–97
Sea bass, 91–92
Seeley, Thomas, 150, 152, 196n33
Segal, David, 183n43
Segal, Mady, 183n43
Selfish gene, 111
Selfish Gene (Dawkins), 111
Selfish teamwork path to cooperation: alarm calling in birds, 123–25, 192n28; bluejay example of, 122–23; as by-product mutualism, 108, 111–14, 128, 131–33, 169–70, 190n7; chimpanzee example of, 116–18; and human cooperation, 109–11, 127–33; lion example of, 21–22, 26, 114–16; and mobbing behavior, 125–27; Pascal's cost/benefit analysis, 109, 110–11; pitfalls associated with, 169–70; and pseudoreciprocity, 190n7; sparrow example of, 120–22; wrasse example of, 118–19; yellowtail example of, 120
Selfishness, 10–12. *See also* Selfish teamwork path to cooperation
Self-sacrificial behavior: of animals, 17–20, 28, 50; of humans, 69, 137
Seltzer, Irwin, 64
Serengeti National Park, 114–16
Sergeant-major damselfish, 119
Serranus tortugarum, 92
Sex. *See* Reproduction
Sherman, Paul, 55
Sierra Club, 67–68
Sinclair, David, 16
Skinner boxes, 122
Skutch, Alexander, 60, 181n24
Slapstick: or Lonesome, No More! (Vonnegut), 63–64
Slaton, Rabbi Eric, 186n14
Smith, Adam, 10–11
Sober, Elliott, 179n6, 198n53
Social contracts, 79–81
Social insects, 48–54, 149–58. *See also* Ants; Bees; Wasps
Sociobiology (Wilson), 35
Soldiers. *See* Military
Soliciting by male baboons, 93–95, 187n23
Sonoran Desert, 22–23, 155–57
Soviet Union. *See* Former Soviet Union
Sparrows, 120–22
Spearnosed bats, 188n29
Speciesism, 173
Spencer, Herbert, 29, 114
Squirrels, 17–20, 165
Stallone, Sylvester, 125
Starlings, 126, 127
State University of New York, 78
Stephens, Dave, 122–23
Stepparents, 65–66
Strand, S., 120
Striped parrotfish, 191n18
Superorganisms, 144, 149–55
Suricata suricata, 181n22

Tachycineta bicolor, 100
Tai National Park, 117–18

Tajfel, Henri, 137–39
Tanagers, 123
Tanzania, 59
Taxes, 64, 67
Teamwork in business, 72–73
Temperature regulation, 152–54
Teresa, Mother, 110
TfT strategy. *See* Tit-for-Tat (TfT) strategy
Thalassoma lucanum, 118–19
Thermoregulation, 152–54
Tit-for-Tat (TfT) strategy, 20–21, 88–90, 92, 98, 100, 104–105, 185–86n12, 186n17
Todes, Daniel, 31–32
Tooby, John, 79–81
"Tragedy of the commons" problem, 175n2
Trait-group selection, 194n16
Tree swallows, 100
Tribal thinking, 162–63
Trinidad, 20–21, 148
Trivers, Robert, 82–83, 98, 184–85nn4–7
Two Treatises of Government (Locke), 8

University of California (Davis), 69
University of Cape Town, 54
University of Chicago, 140–41
University of Michigan, 54
University of Toronto, 88, 101
Urban large-scale gardens, 26–27
Ussher, Bishop James, 172
Utility, definition of, 175n8
Utility customers, 5–6

Vampire bats, 97–99, 165, 188nn26–27
Venezuela, 71–72
Violence. *See* Aggression; Warfare
Visscher, Paul, 52–53
Von Frisch, Karl, 151
Von Neumann, Jon, 84
Vonnegut, Kurt, 63–64

Wade, Michael, 142–43
Wager (Pascal), 110
Waggle dance of bees, 151
Wallace, Alfred Russell, 30–31, 32
Warfare, 95, 101–103
Wason Selection Task Test, 79–81
Wasps, 28, 48
Wealth of Nations (Smith), 10–11
Weiner, Josh, 6
West Indies, 20–21, 148
White-fronted bee-eaters, 61–63, 68–69
Wilkinson, Gerald, 97–99, 188nn26–27
Will, George, 64
Williams, George C., 142, 177n22
Willow tits, 124–25
Wilson, David Sloan, 89, 142–43, 179n6, 198n53
Wilson, E. O., 35, 180n11
Wilson, Margo, 65–66
Wisdom of the Hive (Seeley), 150
Within-group selection, 143
World War I, 101–103
World War II, 103
Worms, 24, 92, 165
Wrasse, 118–19
Wrege, Peter, 61–62
Wright, Robert, 35–36, 178n28
Wright, Sewall, 48, 179n8
Wynne-Edwards, V. C., 141–42

Yanomamo, 71–72
Yeast genome, 16–17
Yellowtails, 120